Progress in IS

Progress in IS encompasses the various areas of Information Systems in theory and practice, presenting cutting-edge advances in the field. It is aimed especially at researchers, doctoral students, and advanced practitioners. The series features both research monographs, edited volumes, and conference proceedings that make substantial contributions to our state of knowledge and handbooks and other edited volumes, in which a team of experts is organized by one or more leading authorities to write individual chapters on various aspects of the topic. Individual volumes in this series are supported by a minimum of two external reviews.

The Series is SCOPUS-indexed.

Adam Wasilewski

Multi-variant User Interfaces in E-commerce

A Practical Approach to UI Personalization

 Springer

Adam Wasilewski
Wrocław University of Science and
Technology
Wrocław, Poland

ISSN 2196-8705   ISSN 2196-8713   (electronic)
Progress in IS
ISBN 978-3-031-67757-1   ISBN 978-3-031-67758-8   (eBook)
https://doi.org/10.1007/978-3-031-67758-8

This Springer imprint is published by the registered company Springer Nature Switzerland AG
The registered company address is: Gewerbestrasse 11, 6330 Cham, Switzerland

If disposing of this product, please recycle the paper.

I dedicate this book to my dad, who has always been there for me and supported me in every moment of my life.

Preface

In the rapidly evolving digital age, e-commerce has witnessed a significant transformation. The interplay between technology and consumer behavior has given rise to various approaches aimed at enhancing the online shopping experience for customers. Among these approaches, the multivariant user interface stands out as a key element for e-commerce companies striving to keep up with current and relevant trends.

The gateway to online shopping has always been the user interface. Initially, it was rudimentary, consisting of simple text-based web pages and clickable links. Users had to navigate through clunky menus, enduring long load times that left them with ambivalent feelings. However, as technology advanced, so did the expectations of online shoppers. Currently, there is an increasing need for engaging, efficient, and personalized experiences. This alteration in user expectations established the groundwork for the progression of e-commerce user interfaces. Additionally, as e-commerce platforms have expanded in both size and sophistication, the importance of user-centric personalization has gained recognition. A multivariant user interface goes beyond merely adapting interfaces to different devices; it represents a strategy for delivering a personalized experience to specific recipients. This transition to personalized tailoring marked a significant juncture in the development of e-commerce. A multivariant user interface allows information, products, and features to be presented in a manner that appeals to different customers. For instance, a visitor to an e-commerce site focusing on fashion may encounter a homepage showcasing popular trends, while a different user with a track record of outdoor gear acquisitions might be greeted with a presentation of new hiking gear. This level of personalization is made possible by algorithms that consider a user's past behavior, preferences, products of interest, and even real-time data.

The incorporation of extensive data analytics has played a crucial role in enabling personalization. In today's e-commerce landscape, platforms collect vast amounts of user interaction data, ranging from viewed products to time spent on separate pages. With the assistance of machine learning algorithms, this data can be analyzed and leveraged to create personalized product suggestions, fine-tune pricing policies, and optimize the overall user experience. Thus, the development of a multivariant user interface in e-commerce is not solely focused on aesthetics; it is about creating an interface that adapts to the distinct requirements and inclinations of each user. Its primary goal is to deliver the right content at the right time and in the right manner, with the flexibility to expand as needs and technical possibilities evolve. One of the fundamental goals of a multi-variant UI is to enhance the user experience. As users increasingly interact with e-commerce platforms, they come to expect a seamless and enjoyable journey. The multivariant UI achieves this by reducing the gap between the user and the desired product. For example, during the checkout process, it allows the system to adapt to the user's preferences, offering a variety of payment options, shipping methods, and even personalized discounts or promotions. This not only streamlines the purchase process but also incorporates a level of personalization capable of satisfying any potential customer.

In summary, the development of multi-variant UIs in e-commerce represents a move away from rigid, one-size-fits-all interfaces toward flexible, customizable, and responsive user experiences. This advancement mirrors the changing demands of online shoppers and the innovative potential of present-day technology.

This publication covers the theoretical foundations, requirements, assumptions, implementation methods, and the results of validating and verifying a platform that comprehensively supports multivariant user interfaces in e-commerce. The research described here primarily focuses on the practical aspects of managing personalized layouts in online shops. The obtained results serve as a basis, but also an incentive, for the wider application of the proposed approach. Implementing a solution to offer dedicated user interfaces in e-commerce is a challenging task. Unlike mechanisms that suggest or advertise products, user interface variants require more extensive customer behavior data collection and a distinct processing approach. This demands a suitable architecture and carefully chosen analytical techniques. The comprehensive handling of this issue was addressed through research and development, and the findings are presented in this publication. As part of the project, an architecture for delivering custom interface variants in e-commerce was developed and practically verified. The results allowed the identification of clustering techniques that produced very good results in a specific business context. The experimental validation of the proposed concept and the tangible benefits of using personalized user interface modifications were also verified.

The issues discussed here hold significant relevance for those involved in developing e-commerce platforms seeking to exploit the opportunities offered by fully personalized user interfaces, surpassing the limitations of conventional solutions. The combination of machine-learning methods and personalization presents substantial opportunities for advancing e-commerce. This publication offers practical insights into the challenges associated with delivering dedicated user interfaces, highlighting risks, potential errors, and improvement opportunities. It provides a technical and business foundation for the widespread implementation of such personalization in e-commerce.

Wrocław, Poland   Adam Wasilewski
April 2024

Acknowledgments

I would like to thank the Fast White Cat S.A. implementation team—Ela, Norbert, Mateusz, Robert, Andrzej, and others involved in the AIM^2 project—for great work in designing, implementing, and verifying the platform for serving multivariant user interfaces in e-commerce. Their knowledge, experience, and patience helped to prepare an innovative solution that can significantly change the perception of UI personalization.

Also, I would like to thank OTCF S.A., and in particular Mr. Tomasz Kaczmarek, Mr. Grzegorz Gracz, and Mr. Waldemar Tabaczynski, for allowing practical verification of the proposed solution and for valuable advice and suggestions during the project.

The research work and the implementation of a platform for serving a multivariant e-commerce user interface were carried out within the project "Self-adaptation of the online store interface for the customer requirements and behavior" cofunded by the National Centre for Research and Development under Sub-Action 1.1.1 of the Operational Program Intelligent Development 2014–2020.

Contents

Acronyms

ABS	Average Basket Size
ABV	Average Basket Value
ASP	Average Selling Price
AI	Artificial Intelligence
AJAX	Asynchronous JavaScript and XML
AOV	Average Order Value
API	Application Programming Interfaces
AR	Augmented Reality
B2B	Business to Business
B2C	Business to Customer
B2E	Business to Employer
B2G	Business to Government
BIRCH	Balanced Iterative Reducing and Clustering using Hierarchies
C2B	Customer to Business
C2C	Customer to Customer
C2G	Customer to Government
CBF	Content-based filtering
CF	Collaborative filtering
CLARA	Clustering Large Applications
CLARANS	Clustering Large Applications based on RANdomized Search
CLIQUE	CLustering In QUEst
CSS	Cascading Style Sheets
COVID-19	COronaVIrus Disease 2019
CR	Conversion Rate
CTR	Clickthrough Rate
CURE	Clustering Using Representatives
CVV	Customer Visit Value
DBSCAN	Density-Based Spatial Clustering of Applications with Noise
DCE	Distributed Computing Environment
DENCLUE	DENsity-based CLUstEring
DenGRID	Density Grid-based Clustering

ERP	Enterprise Resource Planning
EU	European Union
GA	Google Analytics
G2B	Government to Business
G2C	Government to Customer
G2G	Government to Government
GDPR	General Data Protection Regulation
GTM	Google Tag Manager
HDBSCAN	Hierarchical Density-Based Spatial Clustering of Applications with Noise
HTML	HyperText Markup Language
JS	JavaScript
KNN	K-Nearest Neighbor
KPI	Key Performance Indicator
ML	Machine Learning
OECD	Organisation for Economic Cooperation and Development
OPTICS	Ordering Points to Identify the Clustering Structure
OSF	Open Software Foundation
PAM	Partitioning Around Medoids
PCR	Partial Conversion Rate
PHP	PHP: Hypertext Preprocessor former Personal Home Page
PMM	Personalisation Maturity Model
PWA	Progressive Web Application
REST	REpresentational State Transfer
RFM	Recency, Frequency, Monetary
ROCK	RObust Clustering using linKs
SaaS	Software as a Service
SOAP	Simple Object Access Protocol
STAGE	STreaming Algorithm for Grid basEd Clustering
STING	STatistical INformation Grid
TMS	Tag Management System
UI	User Interface
UX	User eXperience
VDBSCAN	Voronoi-Based Density-Based Spatial Clustering of Applications with Noise
VPN	Virtual Private Network
TSNE	T-distributed Stochastic Neighbor Embedding
UMAP	Uniform Manifold Approximation and Projection
W3C	World Wide Web Consortium
WCAG	Web Content Accessibility Guidelines
WFS	Workflow System
WS	Web service
WHO	World Health Organization
WTO	World Trade Organization
WWW	World Wide Web

Chapter 1
Introduction to the Personalization in E-commerce

1.1 E-commerce

Electronic commerce (e-commerce) refers to any economic activity carried out using electronic media. The Organization for Economic Co-operation and Development (OECD) defines e-commerce as *all forms of transactions relating to commercial activities, including both organizations and individuals, that are based upon the processing and transmission of digitized data, including text, sound, and visual images* [1]. In turn, the World Trade Organization (WTO) defines e-commerce as the production, distribution, marketing, and sale or delivery of goods and services by electronic means [2].

E-commerce is commonly linked to diverse online retailers and auction platforms, which permit the buying and selling of tangible goods (e.g., books, cosmetics, and electrical appliances), intangible goods (e.g., e-books and specific kinds of digital software), and services (e.g., access to audiovisual content, education, and training). E-commerce refers to the process of conducting commercial transactions via the Internet or other computer networks, regardless of the tools or means used [3].

E-commerce can be considered multidimensional, and thus various classifications can be made. Nevertheless, the most popular classification of e-commerce is based on the relationship between the seller and the buyer [4]. Assuming that there are two groups in the market—companies (business—B) and individuals (customers—C), the following e-commerce models can be distinguished [5]:

- A company sells and a company buys, referred to as business to business (model B2B).
- A company sells and an individual buys, referred to as business to customer (model B2C).
- An individual sells and an individual buys, referred to as customer to customer (model C2C).

© The Editor(s) (if applicable) and The Author(s), under exclusive license
to Springer Nature Switzerland AG 2024
A. Wasilewski, *Multi-variant User Interfaces in E-commerce*, Progress in IS,
https://doi.org/10.1007/978-3-031-67758-8_1

- An individual sells and a company buys, referred to as customer to business (model C2B).

The **B2B** model pertains to commercial agreements between economic organizations. In this approach, a company provides its products or services to another company. Examples include the supply of materials between companies or the sale of products to wholesalers by a manufacturer.

The **B2C** model refers to the sale of goods or services by companies to consumers. In this approach, businesses directly sell to end consumers through an IT system. Examples include an e-commerce store where clients can view, select, and purchase items online or an auction platform where firms allow individual customers to make purchases.

The **C2C** model pertains to trade among consumers. It is a model whereby consumers sell products or services to other consumers via an e-commerce platform. Examples include an auction platform where users list items that are purchased by other individuals or sales via social media [6].

The **C2B** model refers to commercial transactions in which consumers offer their products or services to companies. Examples include professionals offering freelance services to companies or selling individual collections to companies trading in numismatic or other antique items.

In the context of e-commerce models, the relationship between e-commerce and e-business is a crucial consideration. Although these terms are often used interchangeably, they have distinct meanings. **E-business** encompasses a wider scope than commerce alone, involving various aspects beyond the mere sale of goods and services. This concept can be defined as the use of the Internet to integrate and facilitate business ventures, e-commerce, communication, and collaboration within a company and with its customers, suppliers, and other business partners [7].

When defining e-business, an additional element, government (G), can be introduced to the *Business–Customer relationship*, which adds further models [8]:

- The exchange of information between administrations, *government to government* (model G2G)
- Information flows from business to administration, *business to government* (model B2G)
- Information provided by administration to business, *government to business* (model G2B)
- Information flows from individuals to administration, *customers to government* (model C2G)
- Information made available by the administration to individuals, *government to customers* (model G2C)

Additionally, another aspect of the e-business relationship involves incorporating the employee (E). In practice, the most prevalent linkage involving employees is the **B2E** model. This approach signifies solutions that provide employees with different types of benefits that are financed or co-financed by their employers.

Fig. 1.1 E-business and
e-commerce models

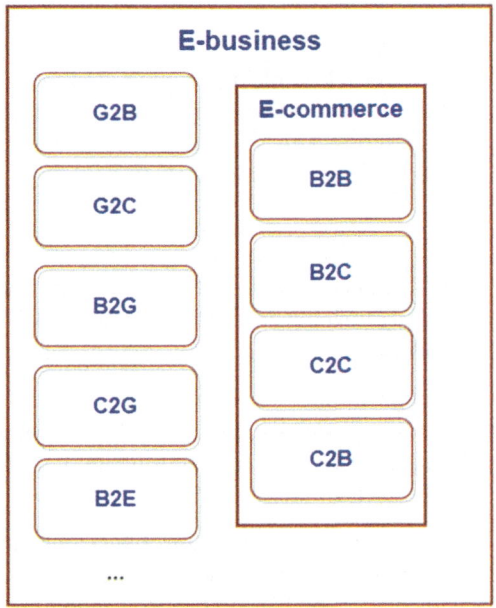

When examining e-business as a whole (Fig. 1.1), e-commerce can be viewed as a component directly related to the buying and selling of products or services. This does not diminish the significance of e-commerce within the broader spectrum of e-business activities. On the contrary, it is the component of e-business that is most accessible to the average Internet user and is frequently utilized by consumers.

E-commerce solutions can be further categorized based on how products or services are offered, distinguishing between:

- Classic online shops (B2C) and wholesale platforms (B2B) are implemented by entrepreneurs using their infrastructure and dedicated software or e-commerce platforms.
- B2C and B2B solutions are offered in the Software as a Service (SaaS) model, which is part of a service provided by specialized companies.
- Marketplaces, auction portals, and stock exchanges serve as sales platforms shared by multiple sellers.
- Tendering platforms are usually dedicated to specific industries or emerging from legal requirements.
- Social media and advertising portals, primarily for small-scale activities and individual sales
- Other sales activities, such as mailing, orders via web forms, and more

This classification highlights the primary approaches to offering goods to consumers. Companies frequently blend various methods to expand their sales channels. For instance, a seller may operate their e-shop while simultaneously showcasing their products on various online marketplaces or auction sites.

The term e-commerce is encompassing and includes a wide range of approaches. For this publication, it is narrowed down to activities specifically related to online shopping on sales platforms, such as various types of online shops. This scope excludes all aspects of sales through other channels, advertising, and online marketing.

Another classification of e-commerce businesses is centered on logistical aspects, particularly how buyers' orders are fulfilled. This classification includes the following models:

- Through an in-house warehouse
- Through a specialized multivendor warehousing company
- Using dropshipping (i.e., fulfillment of deliveries on behalf of the seller from the wholesaler's warehouse) [9]

The strategies for warehouse management described above apply to any e-commerce model where the seller is a commercial entity. In the case of sales by individuals, logistics are typically handled by the seller due to the limited scale of operations. However, there are instances where individual logistics for sales through e-commerce platforms can receive support from the platform owner. This support may include predefined delivery methods and assistance with parcel shipping.

Another way to categorize e-commerce activities is based on the subject of the trade, which can be divided into:

- The sale of tangible (physical) goods
- Sales of intangible goods (especially digital)
- The sale of various types of services

In many cases, a seller's offering may comprise multiple types of goods, for example, selling a timepiece with an engraving dedication service. This division has implications for order fulfillment and logistics, as tangible items are typically available in limited quantities, digital items can often be sold in any quantity (sometimes with a digital signature designated to a specific buyer), and services can be dispensed within a designated time frame.

In conclusion, it is important to note that the classical approach to e-commerce encompasses four key perspectives [10]:

1. The communication perspective—it involves the electronic delivery of information, products, services, or payments.
2. The business process perspective—it leverages technology to automate business transactions and workflows.
3. The service perspective—it facilitates cost reductions while enhancing the speed and quality of service delivery.
4. The online perspective—it focuses on buying and selling products and information online.

This approach highlights that e-commerce extends beyond the actual exchange of products and includes presales and postsales activities throughout the supply chain.

E-commerce has evolved into an integral part of the global economy, fundamentally transforming the way businesses operate and consumers engage in commercial activities. Its significance to the economy cannot be overstated, as it has had and continues to exert a profound influence on various sectors and facets of economic development. Virtually, since its inception, e-business in its broadest sense has experienced dynamic growth, driven in part by technological advancements and the increasingly widespread access to the Internet. E-commerce is reshaping the way businesses conduct transactions, enabling them to reach global audiences and transcend traditional geographical boundaries. The proliferation of online trading platforms and digital services has made it possible to shop quickly and conveniently, eradicating the constraints of time and distance. Consequently, companies of all sizes, from small start-ups to multinationals, have gained access to a vast market, fostering growth and innovation. The dynamism of market change is vividly depicted by the growth in e-commerce's share of retail sales from 2015 to 2020 and forecasts for the years 2021–2024 (Fig. 1.2).

In recent years, the COVID-19 epidemic has provided an additional boost to the growth of e-commerce. The restrictions imposed by the disease have impacted people worldwide, including retailers who had to contend with limitations on the movement of people and customers' fears of crowded shops or shopping centers. Shortly after the outbreak, a survey revealed that 52% of consumers wished to avoid shopping in brick-and-mortar stores and crowded places, with 36% stating they would refrain from in-store shopping until receiving the new vaccine [12]. The traditional approach to shopping had to change virtually overnight, presenting new challenges and opportunities for e-commerce solutions. During the pandemic, many people tried online shopping for the first time [13], resulting in a significant increase

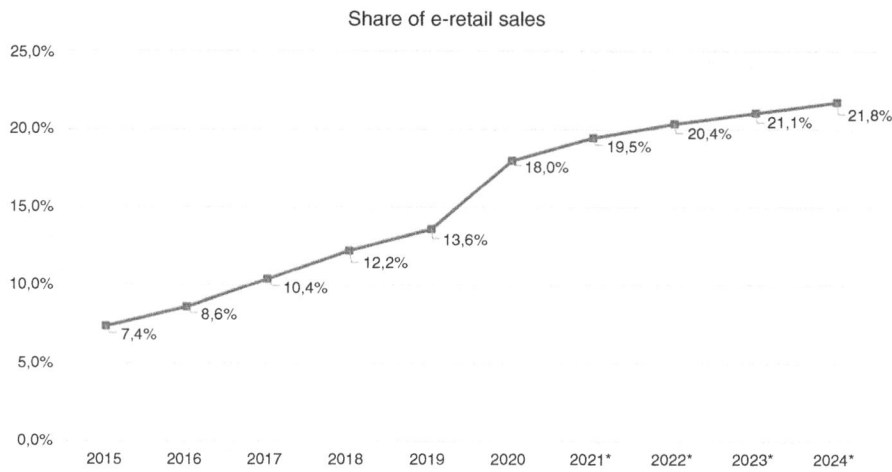

Fig. 1.2 E-commerce share of total global retail sales 2015–2024 [11]

in online sales compared to the period preceding the pandemic. Remarkably, online sales did not experience a significant decline after the pandemic, suggesting that the pandemic has permanently changed consumer behavior [14].

Surviving the pandemic necessitated that companies develop skills in managing relationships with business partners, addressing information asymmetry, improving logistics operations, and enhancing virtual customer relationship management [15]. Many companies had to quickly transition their business to the Internet, often in an ad hoc manner, while seeking alternative ways to reach their customers. Over time, the online channel was sometimes developed and systematized, such as through the implementation of e-shops or by leveraging auction portals and marketplaces. However, it is important to note that for many companies, including those operating online, COVID-19 proved to be an insurmountable challenge, leading to their closure [16].

The COVID-19 pandemic undeniably left a significant impact on electronic business, including e-commerce, from the perspective of both entrepreneurs and individual customers. It compelled companies to accelerate their digital transformation efforts and convinced consumers to shift toward online shopping [17]. One of the noteworthy outcomes of these changes during COVID-19 was the rapid increase in the value of e-commerce in Europe, surging from €633 billion in 2020 to €718 billion in 2021 [18].

However, achieving success in e-commerce is far from straightforward. Amid the various attributes of e-commerce, its competitiveness often takes center stage. The nature of e-commerce makes it easy for potential customers to access multiple offers from competing sellers. Companies employ a range of tactics to entice customers to purchase their quest for a competitive edge. Common strategies include competing on factors such as price, product/service quality and desirability, accessibility, or support. Increasingly, customers are drawn by tailored communication that caters to their specific needs and requirements.

1.2 On the Way to Personalization

Personalization, defined by the Oxford Dictionary as the action of designing or producing something that meets someone's requirements, ranks among the most widely used marketing strategies. Its roots can be traced back to the first personalized direct marketing letters in the 1870s [19]. On a broader scale, personalization has been employed in marketing since the 1950s [20]. With the development of the Internet and information technology in the 1990s, personalization found its application in electronically delivered content and became an integral part of direct marketing.

A cross-sectional survey of scientific articles pertaining to the concept of personalization [21] identified the timing of the appearance of ICT-related topics in these publications:

- Electronic commerce (around 2007)
- Internet marketing (around 2009)

- Recommendation systems (around 2012)
- Web personalization (around 2013)
- Machine learning (around 2018)
- Artificial intelligence (around 2019)

Presently, personalization is acknowledged as one of the key trends in e-commerce development. Effective personalization can have a positive impact on customer satisfaction and, consequently, their willingness to make purchases [22]. Furthermore, it can generate profit for manufacturers, retailers, or service providers while simultaneously adding value for the consumer.

In the literature, various definitions of personalization can be found, and three of these are particularly noteworthy in the context of the issues discussed below [21, 23].

The first definition portrays personalization as a process capable of modifying various aspects of a system (such as functionality, interface, information content, or uniqueness) to increase its relevance to the user [24].

The second definition delineates two primary approaches to personalization: user-driven adaptivity and automatic adaptivity [25]. An additional option, expert-led adaptivity, is also viable. In this scenario, personalization is entrusted to an expert who determines how and to what extent personalization should occur, based on an analysis of the collected behavioral and contextual data.

The third definition is framed around three questions: *Who* performs the personalization (a customer or a system), *whom* it is targeted at (an individual or a group), and *what* is personalized (features, content, user interface, access, channel, promotions, etc.) [26].

However, most definitions converge on the primary objective of personalization, which is to provide customers with dedicated, engaging, and satisfying content, with the assumption that this will lead to improved sales results. It can be viewed as a whole process aimed at creating relevant, tailored interactions to enhance the customer experience. The main data sources in this process are the unique behavioral data of the recipient, as well as the behavioral data of similar individuals [27].

The personalization process involves several steps that are used to tailor products, services, or experiences to the individual needs and preferences of customers or other types of recipients. The number of steps in the process depends on the approach taken, but the four main ones are:

1. Data collection
2. Data analysis
3. Solution design
4. Solution implementation

Such an approach is presented by the **4D** (Data, Decisioning, Design, and Distribution) framework proposed by McKinsey & Company (Fig. 1.3).

Data foundation serves as a repository of information about customers, gathered from both internal sources within the organization and external sources, including

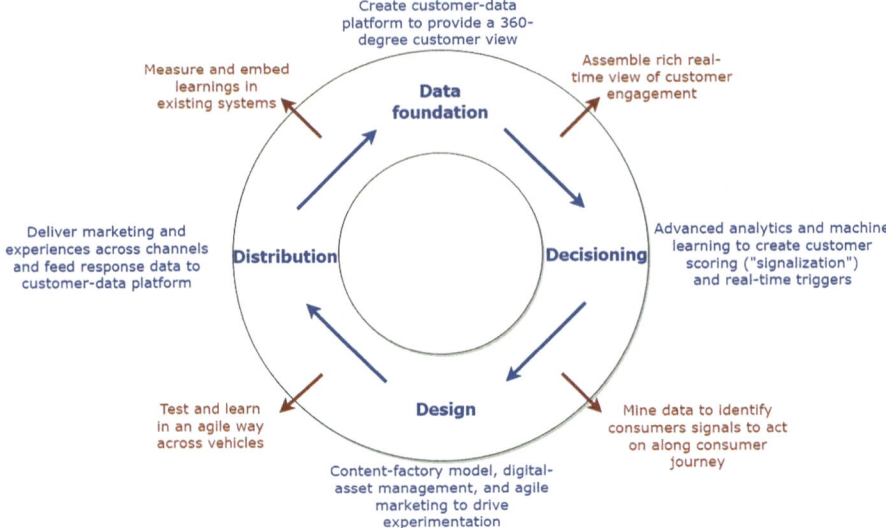

Fig. 1.3 Personalization process framework [28, 29]

social media. Its purpose is to collect data about potential recipients of personalized communication, including their activities, motivations, expectations, and the context in which they operate. An equally significant source of data is the feedback generated by the implemented personalization solutions, closing the feedback loop and enabling the continuous improvement of recommendations.

Decisioning encompasses a set of tools and techniques specifically designed to analyze data and provide recommendations on how to personalize communications to the customer. This may involve a combination of business rules, regression models, machine learning algorithms, and more, indicating *what, when, how,* and *where* to deliver content to customers to enhance their experience. The outcome of this analysis can include signals, which can be basic, like identifying items browsed but not purchased, or more advanced, such as segmenting customers based on various characteristics that describe them. The data analysis methods employed and the signals generated are contingent on the level of *personalization maturity* adopted by the organization.

Design involves the creation of broadly defined content intended for target customers. This content can take various forms, such as articles and publications tailored to the customer's past behavior or decisions, product recommendations, individual promotions, prices, and more. Personalization should not only focus on the content itself but also consider the structure and design of the message.

Distribution entails delivering prepared offers through selected channels. Traditional distribution channels include mail, radio, and television. Today, electronic channels such as email, social media, and e-commerce platforms are widely utilized. It is crucial to note that the personalization process does not conclude with the

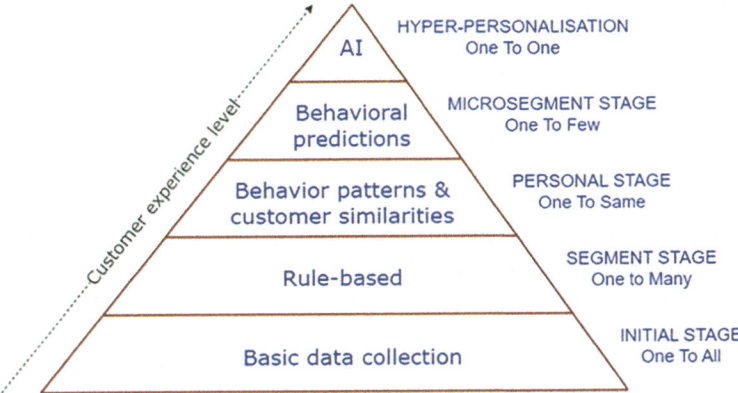

Fig. 1.4 Personalization maturity model [30]

delivery of the customized message; it also encompasses the collection of feedback to gauge the customer's response to the personalized content. By analyzing the impact of personalized communication on customer behavior and decisions, tailored offers can be optimized in subsequent iterations of the process.

The organization's ability to leverage the opportunity to deliver personalized content is closely tied to its **personalization maturity**. This maturity level determines the sophistication of offer creation and distribution. The Personalization Maturity Model (PMM) serves as a tool to assess an organization's current state of personalization activities (Fig. 1.4).

The PMM consists of five levels, each defined by the size of the set of customers receiving a variation of personalized content:

- **One to All**: This is the basic development phase of personalization, involving the creation of customer information. In this phase, all recipients receive the same communication.
- **One to Many**: At this stage, customer segments are created based on collected data, and specific content is provided to these segments based on selected differentiating characteristics, which may include geographic and demographic factors.
- **One to Some**: This phase involves deeper customer segmentation, taking into account behavioral patterns, leading to the division into more groups based on similarities and differences.
- **One to Few**: In this stage, customer segmentation deepens further, considering psychological data, such as buying motivations [31] in addition to geographical, demographic, and behavioral information.
- **One to One**: The final phase, referred to as hyper-personalization, entails the creation of individual, personalized customer experiences. Artificial intelligence [32] is employed to achieve the best possible predictive personalization for each customer [33].

While hyper-personalization may seem like an ideal response to individual customer expectations, its widespread adoption faces technological and legal barriers. It demands the processing of huge amounts of data and maintaining individualized content for each customer, as well as the ability to make inferences based on limited customer information. In practice, it has a chance to be implemented in the case of systems that have a small and finite number of users (e.g., ERP systems, WFS, etc.) [34]. This is compounded by growing customer concerns about privacy and the increasing importance of data protection as a top priority in consumer interactions [35]. Aligned with current trends, companies are shifting away from sourcing data from external sources and concentrating on self-acquired data (*zero-party* and *first-party*) with established privacy policies. However, this trend results in limited customer information, making high-level personalization (and hyper-personalization) potentially misplaced and even counterproductive. As a result, many companies do not advance beyond phase 3 (*One to Some*) in their personalization efforts. More advanced personalization is usually used by companies offering highly exclusive offers to select customers [36].

Nevertheless, personalization in e-commerce, primarily at level 2 or 3 of the maturity model, is currently the market solution that customers desire, as evident in various studies. One study shows that 76% of customers consider receiving personalized communications a key factor in their willingness to purchase products from a new brand, while 78% believe that personalized content increases the likelihood of repeat purchases [37]. According to other surveys, the most common expectations from e-shop customers include personalized navigation (75%), relevant product or service recommendations (67%), personalized communication (66%), targeted promotions (65%), and celebrating milestones achieved (61%) [38].

Personalization processes are now commonly supported by information systems known as *personalization* engines. These engines serve three primary use cases [27]:

- Digital commerce: Tailoring communication and shopping experiences across digital channels, such as websites and mobile devices
- Marketing: Adapting the content of marketing campaigns, messages, and engagement across marketing and communication channels, including websites, email, search, advertising, and mobile
- Customer experience: Customizing online and offline experiences to enhance customer satisfaction, loyalty, and advocacy, which can include chatbots, digital kiosks, voice assistants, and augmented reality (AR)

It is worth noting that the majority of personalization activities in electronic channels, including e-commerce, focus on delivering personalized content, without supporting dedicated user interface design. While customers' needs and preferences vary, the user interface in most e-shops remains the same for all users. The complexity of interface personalization is higher than that of distributing personalized messages, advertisements, product recommendations, or prices. Implementing a dedicated UI requires the technical capability to modify the layout of the e-shop dynamically, which is more challenging than simply displaying a product or advertisement in a predetermined location [39]. The possibilities for personalizing e-commerce interfaces depend on the general capabilities of personalizing web interfaces, given that e-commerce is fundamentally based on web solutions.

1.3 Personalization of Web-Based User Interfaces

The origins of web interfaces can be traced back to the early 1990s when Tim Berners-Lee and Robert Cailliau introduced their World Wide Web (WWW) Hypertext project [40]. Simultaneously, Berners-Lee launched the first web server and developed the first web browser. This groundbreaking solution made its web debut on August 6, 1991, marking the inception of web user interfaces. Over the years, software applications, including servers and browsers, have evolved. The text-based browsers in use during the mid-1990s (such as Lynx) were swiftly replaced by "window-based" solutions, like Mosaic in 1992 and Netscape Navigator in 1994. Nevertheless, web pages continue to be constructed using the same language, HTML (HyperText Markup Language), albeit with modifications and extensions, which has been in use practically since the inception of the WWW.

HTML is built on the concept of text enhanced with links pointing to other text within the same document or to external documents stored elsewhere [41]. It provides the structure and content of a web page using a predefined set of tags and attributes. HTML is generally a static language, meaning that the content and layout of a web page remain unchanged unless the HTML code is manually modified. However, the appeal of web-based systems can be heightened by incorporating dynamic elements that influence the content and presentation of the information provided [42].

Technologies for delivering dynamic web content have evolved over the years. Today, system designers have a variety of solutions at their disposal to prepare and deliver dynamic information or forms in conjunction with HTML. These solutions differ in terms of where data processing occurs, which can take place either on the client application (browser) side or on the server side.

Commonly used client-side solutions include:

- *Cascading Style Sheets* (CSS): This is used to style and format HTML elements by controlling the appearance of various components, such as fonts, colors, layouts, and animations.
- *JavaScript* (JS): This is a scripting programming language that enables client-side interactivity. It allows for the manipulation of HTML elements, handling user interactions (such as form filling and submission, button clicks), asynchronously retrieving and displaying data from servers using AJAX technology, and dynamically updating web page content.
- *Asynchronous JavaScript and XML* (AJAX) [43]: This technology enables websites to asynchronously update and retrieve data from servers without completely reloading the page, thus maintaining a seamless user experience.

Server-side solutions can be differentiated as follows:

- *Programming languages* (e.g., PHP [44], Python [45], and Ruby [46]): These languages are used for data processing and server-side logic. They enable the dynamic generation of HTML, communication with users and databases, performing calculations, and ultimately serving data to the browser.
- *Development platforms* (e.g., Node.js) [47]: Node.js is a back end JavaScript runtime environment that allows JavaScript code to be executed outside of a web browser.
- *Web service* (WS): These are software systems designed to support interoperable machine-to-machine interaction over a network [48], with popular approaches including Representational State Transfer (REST) [49] and Simple Object Access Protocol (SOAP) [50].
- *Application programming interfaces* (APIs): APIs allow web applications to communicate and exchange data with external services or systems. This can involve using solutions such as web services or database queries to display obtained information in the browser dynamically.

Personalizing the web interface requires user identification, in addition to the ability to provide dynamic content. This function is often facilitated by what are known as cookies—small blocks of data generated by a server while a user is browsing a website and placed on the user's computer or other device by the user's web browser. The *cookie* mechanism was initially developed by Netscape Communications and was first employed in the Netscape Navigator v.2 browser [51]. Its original specification was described in RFC 2109 [52], then superseded by RFC 2965 in October 2000 [53], and finally written in RFC 6265 in 2011 [54].

A *cookie* allows the storage of name, value, and attributes (in name/value pairs) on the client application side. These attributes can include information such as the cookie's expiry date, domain, and flags (e.g., secure and HttpOnly) [55]. Despite its relatively simple structure, this mechanism is a crucial component of identity management in web-based systems and serves various applications, including session management, tracking, and personalization.

In the context of **session management**, cookies are primarily utilized to store information about the user during their session on a website. This allows users to return to the same page they were previously logged into without having to log in again for the duration of the cookie's validity, as the authorization information is retained in this file. In the case of e-commerce, cookies are often employed to store the identification of the shopping basket in the e-shop. This enables customers to continue shopping after a while without losing the contents of their shopping basket, which is stored in the shop's database and linked to the identifier stored in the cookie.

Cookies are also used to **track** the activity of web users. After a cookie is stored on the first visit, the browser automatically sends information to the server each time a new page is requested from the site. The server may store the URL of the page requested, the date/time of the request, and information about the identification of the associated file. By analyzing this data, it is possible to determine which pages the user has visited, in what order, and for how long. In the context of e-commerce, the tracking of users through the use of cookies makes it possible to gather information about purchasing habits and, consequently, to tailor the content sent to customers.

Cookies can be used for **personalization** in several ways. In the simplest case, they are used to store data (such as email address, shipping address, etc.) or preferences selected by the customer (such as avatar, background color, etc.). More advanced solutions can process information obtained from tracking customer history and generate personalized recommendations for products, ads, and promotions.

The use of cookies offers many benefits to both parties and allows companies to get to know their customers. It makes it possible to identify the customer, his/her behavior, habits, and preferences and thus personalize the content offered. On the other hand, it poses a threat to privacy and can lead to abuse. For this reason, steps have been taken in recent years to regulate the tracking of Internet users, including restrictions on the use of cookies, which will be discussed in the next chapter.

In general, in web-based systems, the modification of **content**, **structure**, and **presentation** can be applied, either for personalizing selected aspects or for making comprehensive changes. For systems focused on information distribution, such as web portals, content modification is often the most relevant. In systems that support learning, structural modification is commonly used to alter the order in which materials are presented. Changing the way information is presented can be effectively employed in various types of web-based systems [42].

Content customization may encompass the following [56]:

- Additional (optional) explanations are displayed only to selected users (e.g., those visiting the site for the first time).
- Additional detailed information
- Personalized recommendations on various aspects of social and economic life, allowing information to be filtered based on user-specified preferences or identified by system solutions
- Contextual hints that respond to the interaction context are derived from the user model and the current cursor position. These hints can highlight specific sections or display additional information in new windows.
- Adjustments to language, currency, and units of measurement, tailored to the context of the website user

The structure of a web system is defined by hypertext (hypermedia) links, which can be categorized as contextual (occurring within the presented content) and context-free (independent of the content, such as buttons). Consequently, the adaptation of the structure may include the following [56]:

- Derivative adaptation, which results from adapting the content, can involve actions like removing a link.
- Link sorting, often found in recommendation systems, is used for sorting links based on the user model in online stores and marketplaces.

- Marking of links, whether contextual or out of context, based on the user's activity history (e.g., visit history) or recommending actions to the user (e.g., suggesting pages to visit or buttons to click)
- Hiding and showing links based on various factors such as the customer's history (e.g., hiding sections for new customers for returning customers), preferences (e.g., enabling or disabling sections of the service), status (e.g., showing sections for logged-in users), or generated recommendations (e.g., displaying links to recommended products)

Presentation changes do not alter the content and structure of the site. They primarily involve modifications to the visual elements, such as graphics, text, or audio. Common changes in this category include adjustments to color schemes, font type and size, and the placement of content.

When examining the technological aspects of personalizing web interfaces, including those in e-commerce solutions, it becomes evident that the implementation of this concept can be achieved through various mechanisms. However, alongside technological considerations, privacy issues can exert a significant impact on the extent of personalization.

1.4   Privacy Challenges in Personalization

The growing economic importance and widespread use of online shopping opportunities are compelling policymakers to introduce regulations that establish the framework for the operation of all e-commerce participants. These regulations are primarily implemented within the European Union (EU), but some are also adopted in other markets as best practices or as prerequisites for trading with the EU. Some of the well-known rules are directly related to the purchasing process, such as the right of withdrawal within 14 days of purchase (Directive 2011/83/EU [57]) and the obligation to inform about the lowest price before a promotion (the so-called *the Omnibus Directive*—Directive 2019/2161/EU [58]). However, from a personalization perspective, two other sets of regulations have a significant impact on the ability to collect and process customer data for delivering specific content. The first set relates to the general principles governing the processing of personal data, and the second set concerns the conditions governing the use of cookies.

The General Data Protection Regulation (GDPR) is a comprehensive data protection law that was implemented in the European Union (EU) on May 25, 2018 as Regulation 2016/679/EU [59]. Its main objective is to strengthen and harmonize data protection for all individuals in the EU and provide them with more control over their data. The GDPR establishes a set of rights and principles related to the collection, use, storage, and disclosure of personal data. Some of these principles significantly affect the ability to collect and process customer data in e-commerce, impacting the capacity to generate recommendations for personalization. Key GDPR principles that influence personalization capabilities include:

- *Purpose limitation.* The collection of personal data requires the declaration of purposes for data collection, and customers must be informed about them. The data collected may only be used for the specified purposes and by the legal basis for data processing under the GDPR.
- *Data minimization.* Organizations should only collect and process personal data that is necessary to achieve the stated purposes. Unnecessary or excessive data should not be collected or stored.
- *Storage limitation.* Personal data should not be retained in a form that allows individuals to be identified for longer than necessary for the specified purposes.

In addition, the GDPR grants consumers additional rights, including the right to access their data, the right to rectification, the right to erasure (*right to be forgotten*), the right to object to processing, the right to data portability, and the right not to be subject to automated decision-making.

All of these limitations mean that companies aiming to personalize the content they deliver to their customers must exercise caution when collecting information that could be considered personal data.

On the other hand, it should be noted that the GDPR only pertains to personal data, which is defined as *any information related to an identified or identifiable natural person (data subject); an identifiable natural person can be directly or indirectly identified, especially by reference to an identifier such as a name, an identification number, location data, an online identifier, or one or more factors specific to the physical, physiological, genetic, mental, economic, cultural, or social identity of that natural person* [59]. This means that collecting consumer data for personalization purposes is possible but requires careful consideration, compliance with legal requirements, and obtaining consent to collect and process data [60]. Additionally, data analysis should involve processes that utilize anonymization and pseudonymization when processing customer data [61].

On the other hand, regulations strengthen customer confidence and can create long-term value [62]. It is worth noting that the rules outlined in the GDPR only apply to the rights of individuals (consumers) interacting with organizations. Therefore, they do not extend to relationships where the consumer is not present (e.g., B2B) or where the consumer does not engage with the organization (e.g., C2C). Hence, they do not apply to all e-commerce scenarios.

Directive on Privacy and Electronic Communications (e-Privacy Directive) was introduced in 2002 [63]. This regulation required end users to give their consent before information could be stored or accessed on their devices using cookies or similar technologies. The storage of technically unnecessary data on a user's computer could only occur if the user was provided with information about how the data would be used and had the opportunity to opt out. These restrictions did not apply to necessary cookies to provide a service, such as maintaining settings or remembering the contents of a shopping basket in an e-shop. In 2009, new regulations were introduced, replacing the requirement to allow customers to reject cookies with a requirement to obtain consent to store cookies [64]. Furthermore, the information contained in many cookies can be considered personal data under the GDPR and requires a legal basis for processing. Given this, it is likely that the collection and processing of data for personalization and behavioral advertising purposes should be preceded by obtaining appropriate consent from the end user [65].

The above EU regulations do not preclude the use of cookies to collect information about customers, their actions, and choices, but their use is subject to several conditions [66]. Consent to cookies, including those for potential personalization processing, must be given voluntarily and unambiguously, cannot be obtained by default through prechecked boxes, and must be as easy to refuse as it is to give. Additionally, consent must be informed and specific, relating to the precise purposes for which the data will be used, and the organizations that will use the consent must be identified in the body of the consent.

The use of cookies faces more than just legal barriers. For several years, browser vendors have also been imposing restrictions on the use of so-called third-party cookies, which are considered a threat to users' privacy and anonymity. Third-party cookies are generated and placed on a user's device by a site other than the one being visited, often through elements like images or advertisements [67]. In practice, these cookies can track a user's browsing history, are used for displaying targeted advertising, and can potentially be used to collect data for personalization purposes. Most modern browsers now allow users to configure their privacy preferences to block third-party cookies. In some cases, such as Safari and Firefox, blocking these cookies has been the default setting since 2020. The vendor of one of the most popular browsers, Chrome, plans to start blocking such cookies by default at the end of 2024 [68].

The various barriers to the use of cookies certainly limit the ability to obtain information about customers and their online behavior. From a privacy perspective, the actions taken by regulators and browser vendors are positive for users but undoubtedly affect the potential personalization of content delivered to them. The answer is for companies to move away from sourcing data from external sources to sourcing data themselves within implemented privacy policies [30]. While these approaches only allow the user to be known in terms of their use of a particular website, they guarantee their privacy and in the long term may persuade them to agree to personalization.

References

1. OECD (1997) Sacher report, OECD digital economy papers, No. 29. OECD Publishing, Paris. https://doi.org/10.1787/237058611046
2. WTO (1998) Work programme on electronic commerce. World Trade Organization, Geneva, WT/L0274, Geneva. https://www.wto.org/english/tratop_e/ecom_e/wkprog_e.htm. Cited 15 Jul 2023
3. Skorupska J (2017) E-commerce. Strategia – Zarzadzanie – Finanse. PWN, Warszawa

4. Chaffey D (2011) E-business & e-commerce management: strategy, implementation and practice. Pearson, New Jersey

5. Turban E, King D, Lee JK, Liang T-P, Turban DC (2015) Electronic commerce: a managerial perspective. Pearson, New Jersey

6. Huang Z, Benyoucef M (2013) From e-commerce to social commerce: a close look at design feature. Electron Commer Res Appl. https://doi.org/10.1007/s001090000086

7. Combe C (2006) Introduction to e-business: management and strategy. Routledge, London

8. Chaffey D (2015) Digital business and e-commerce management. Pearson, New Jersey

9. Wasilewski A (2018) Dropshipping w e-commerce. In: Wspomaganie zarzadzania z wykorzystaniem technologii IT, Wydawnictwo Politechniki Czestochowskiej

10. Kalakota R, Whinston A (1997) Electronic commerce: a manager's guide. Addison-Wesley, Reading

11. International Trade Administration (2021) Impact of COVID Pandemic on eCommerce. https://www.trade.gov/impact-covid-pandemic-ecommerce. Cited 15 Jul 2023

12. Bhatti A, Akram H, Basit H, Khan A, Mahwish S, Naqvi R, Bilal M (2020) E-commerce trends during COVID-19 Pandemic. Int J Future Gener Commun Netw 13:1449–1452

13. Ecommerce Europe (2021) European E-commerce Report. Ecommerce Europe. https://ecommerce-europe.eu/wp-content/uploads/2021/09/2021-European-E-commerce-Report-LIGHT-VERSION.pdf. Cited 15 Jul 2023

14. Chen Y, Hou M, Lou Y, Zhao Y (2022) The impact of the epidemic on e-commerce industry. In: Advances in economics, business and management research. https://doi.org/10.2991/aebmr.k.220307.360

15. Rapaccini M, Saccani N, Kowalkowski C, Paiola M, Adrodegari F (2020) Navigating disruptive crises through service-led growth: The impact of COVID-19 on Italian manufacturing firms. Ind Mark Manag. https://doi.org/10.1016/j.indmarman.2020.05.017

16. Zhao X (2022) Analysis of the impact of the COVID-19 on E-commerce In: 4th international conference on global economy and business management (GEBM 2022), vol. 23. https://doi.org/10.54691/bcpbm.v23i.1473

17. Matarazzo M, Penco L, Profumo G, Quaglia R (2021) Digital transformation and customer value creation in Made in Italy SMEs: a dynamic capabilities perspective. J Bus Res. https://doi.org/10.1016/j.jbusres.2020.10.033

18. Ecommerce Europe (2022) European E-commerce Report. Ecommerce Europe. https://ecommerce-europe.eu/wp-content/uploads/2022/06/CMI2022_FullVersion_LIGHT_v2.pdf. Cited 15 Jul 2023

19. Vesanen J (2007) What is personalization? A conceptual framework. Eur J Mark. https://doi.org/10.1108/03090560710737534

20. Bauer CL, Miglautsch J (1992) A conceptual definition of direct marketing. J Direct Mark. https://doi.org/10.1002/dir.4000060204

21. Chandra S, Verma S, Lim WM, Kumar S, Donthu N (2022) Personalization in personalized marketing: trends and ways forward. Psychol Mark. https://doi.org/10.1002/mar.21670

22. Desai D (2016) A study of personalization effect on users' satisfaction with e-commerce websites. J Manag Res 6(2):51–62

23. Montgomery AL, Smith MD (2009) Prospects for Personalization on the Internet. J Interact Mark. https://doi.org/10.1016/j.intmar.2009.02.0

24. Blom JO, Monk AF (2023) Theory of personalization of appearance: why users personalize their PCs and mobile phones. Hum Comput Interact. https://doi.org/10.1207/S15327051HCI1803_1

25. Frias-Martinez E, Chen SY, Liu X (2009) Evaluation of a personalized digital library based on cognitive styles: adaptivity vs. adaptability. Int J Inf Manag. https://doi.org/10.1016/j.ijinfomgt.2008.01.012

26. Fan H, Poole MS (2006) What is personalization? Perspectives on the design and implementation of personalization in information systems. J Organ Comput Electron Commer. https://doi.org/10.1080/10919392.2006.9681199

27. Polk J, McNellis J, Tassin C (2020) Gartner magic quadrant for personalization engines. Gartner
28. Boudet J, Gregg B, Heller J, Tufft C (2017) The heartbeat of modern marketing: data activation & personalization. McKinsey&Company. https://www.mckinsey.com/capabilities/growth-marketing-and-sales/our-insights/the-heartbeat-of-modern-marketing#/. Cited 15 Jul 2023
29. Flavin S, Heller J (2019) A technology blueprint for personalization at scale. McKinsey&Company. https://www.mckinsey.com/capabilities/growth-marketing-and-sales/our-insights/a-technology-blueprint-for-personalization-at-scale#/. Cited 15 Jul 2023
30. Artug E (2023) The ultimate guide to getting started with personalization [in 2023]. Ninetailed. https://ninetailed.io/blog/personalization-guide/. Cited 15 Jul 2023
31. Pappas IO, Kourouthanassis PE, Giannakos MN, Lekakos G (2017) The interplay of online shopping motivations and experiential factors on personalized e-commerce: a complexity theory approach. Telemat Inform. https://doi.org/10.1016/j.tele.2016.08.021
32. Kumar V, Rajan B, Venkatesan R, Lecinski J (2019) Understanding the role of artificial intelligence in personalized engagement marketing. Calif Manag Rev. https://doi.org/10.1177/0008125619859317
33. Kotras B (2020) Mass personalization: Predictive marketing algorithms and the reshaping of consumer knowledge. Big Data Soc. https://doi.org/10.1177/2053951720951581
34. Smereka M, Kolaczek G, Sobecki J, Wasilewski A (2023) Adaptive user interface for workflow-ERP system. Proc Comput Sci. https://doi.org/10.1016/j.procs.2023.10.229
35. Cai H, Mardani A (2023) Research on the impact of consumer privacy and intelligent personalization technology on purchase resistance. J Bus Res. https://doi.org/10.1016/j.jbusres.2023.113811
36. Rahman MS, Bag S, Hossain MA, Fattah FAMA, Gani MO, Rana NP (2023) The new wave of AI-powered luxury brands online shopping experience: the role of digital multisensory cues and customers' engagement. J Retail Consum Serv. https://doi.org/10.1016/j.jretconser.2023.103273
37. Birch K (2022) McKinsey: prioritise personalisation for 10–15% revenue lift. McKinsey&Company. https://businesschief.com/technology-and-ai/mckinsey-prioritise-personalisation-for-10-15-revenue-lift. Cited 15 Jul 2023
38. Arora N, Ensslen D, Fiedler L, Liu WW, Robinson K, Stein E, Schüler G (2021) The value of getting personalization right—or wrong—is multiplying. McKinsey&Company. https://www.mckinsey.com/capabilities/growth-marketing-and-sales/our-insights/the-value-of-getting-personalization-right-or-wrong-is-multiplying#/. Cited 15 Jul 2023
39. Wasilewski A, Kolaczek G (2024) One size does not fit all: multivariant user interface personalization in e-commerce. IEEE Access. https://doi.org/10.1109/ACCESS.2024.3398192
40. Berners-Lee T, Cailliau R (1992) World-wide web. In: Computing in high energy physics. Annecy, France
41. Chapman N, Chapman J (2004) Digital multimedia. John Wiley & Sons, London
42. Sobecki J (2009) Rekomendacja interfejsu użytkownika w adaptacyjnych webowych systemach informacyjnych. Wrocław: Oficyna Wydawnicza Politechniki Wrocławskiej
43. Garrett JJ (2007) Ajax: a new approach to web application. https://courses.cs.washington.edu/courses/cse490h/07sp/readings/ajax_adaptive_path.pdf. Cited 15 Jul 2023
44. Siame A, Kunda D (2017) Evolution of PHP applications: a systematic literature review. Int J Recent Contrib Eng Sci. IT (iJES). https://doi.org/10.3991/ijes.v5i1.6437
45. Oliphant T (2007) Python for scientific computing. Comput Sci Eng. https://doi.org/10.1109/MCSE.2007.58
46. Cooper P (2009) Beginning ruby: from novice to professional. Apress
47. Kaimer F, Brune P (2018) Return of the JS: towards a node.js-based software architecture for combined CMS/CRM applications. Proc Comput Sci. https://doi.org/10.1016/j.procs.2018.10.143
48. W3C (2004) Web services glossary. https://www.w3.org/TR/2004/NOTE-ws-gloss-20040211/#webservice. Cited 15 Jul 2023

49. Gupta L (2004) What is REST. https://restfulapi.net/. Cited 15 Jul 2023
50. W3C (2007) SOAP version 1.2 Part 1: messaging framework, 2nd edn. https://www.w3.org/TR/soap12-part1/. Cited 15 Jul 2023
51. Knuckles C D (2004) Web applications. John Wiley & Sons, London
52. Kristol D, Montulli L (1997) RFC2109: HTTP state management mechanism. https://www.rfc-editor.org/rfc/rfc2109. Cited 15 Jul 2023
53. Kristol D, Montulli L (2000) RFC2965: HTTP state management mechanism. https://www.rfc-editor.org/rfc/rfc2965. Cited 15 Jul 2023
54. Berth A (2011) RFC6265: HTTP state management mechanism. https://www.rfc-editor.org/rfc/rfc6265. Cited 15 Jul 2023
55. Lee W-B, Chen H-B, Chang S-S, Chen T-H (2018) Secure and efficient protection for HTTP cookies with self-verification. Int J Commun Syst. https://doi.org/10.1002/dac.3857
56. Kobsa A, Koenemann J, Pohl W (2001) Personalized hypermedia presentation techniques for improving online customer relationships. Knowl Eng Rev. https://doi.org/10.1017/S0269888901000108
57. European Parliament and of the Council (2011) Directive 2011/83/EU of the European Parliament and of the Council of 25 October 2011 on consumer rights. https://eur-lex.europa.eu/legal-content/EN/TXT/?uri=CELEX:32011L0083. Cited 15 Jul 2023
58. European Parliament and of the Council (2019) Directive (EU) 2019/2161 of the European Parliament and of the Council of 27 November 2019 amending Council Directive 93/13/EEC and Directives 98/6/EC, 2005/29/EC and 2011/83/EU. https://eur-lex.europa.eu/eli/dir/2019/2161/oj. Cited 15 Jul 2023
59. European Parliament and of the Council (2016) Regulation (EU) 2016/679 of the European Parliament and of the Council of 27 April 2016 on the protection of natural persons with regard to the processing of personal data and on the free movement of such data, and repealing Directive 95/46/EC. https://eur-lex.europa.eu/legal-content/EN/TXT/?uri=CELEX%3A02016R0679-20160504. Cited 15 Jul 2023
60. Haddara M, Salazar A, Langseth M (2023) Exploring the impact of GDPR on big data analytics operations in the e-commerce industry. Proc Comput Sci. https://doi.org/10.1016/j.procs.2023.01.350
61. Weitzenboeck E, Lison P, Cyndecka M, Langford M (2022) The GDPR and unstructured data: is anonymization possible? Int Data Priv Law. https://doi.org/10.1093/idpl/ipac008
62. Gruschka N, Mavroeidis V, Vishi K, Jensen M (2018) Privacy issues and data protection in big data: a case study analysis under GDPR. In: 2018 IEEE international conference on big data (Big Data)
63. European Parliament and of the Council (2002) Directive 2002/58/EC of the European Parliament and of the Council of 12 July 2002 concerning the processing of personal data and the protection of privacy in the electronic communications sector. https://eur-lex.europa.eu/legal-content/EN/ALL/?uri=CELEX%3A32002L0058. Cited 15 Jul 2023
64. European Parliament and of the Council (2009) Directive 2009/136/EC of the European Parliament and of the Council of 25 November 2009 amending Directive 2002/22/EC on universal service and users' rights relating to electronic communications networks and services. https://eur-lex.europa.eu/legal-content/EN/TXT/?uri=celex%3A32009L0136. Cited 15 Jul 2023
65. Zuiderveen Borgesius F J (2015) Personal data processing for behavioural targeting: which legal basis? Int Data Priv Law. https://doi.org/10.1093/idpl/ipv011
66. Nouwens M, Liccardi I, Veale M, Karger D, Kagal L (2020) Dark patterns after the GDPR: scraping consent pop-ups and demonstrating their influence. In: CHI '20: proceedings of the 2020 CHI conference on human factors in computing systems. https://doi.org/10.1145/3313831.3376321
67. Mitchell I D (2012) Third-party tracking cookies and data privacy. SSRN Electron J. https://doi.org/10.2139/ssrn.2058326
68. Business Standard (2022) Google now delays blocking 3rd-party cookies in Chrome to late 2024. https://www.business-standard.com/article/technology/google-now-delays-blocking-3rd-party-cookies-in-chrome-to-late-2024-122072800244_1.html. Cited 15 Jul 2023

Chapter 2
Recommendation System for Multivariant E-Commerce Interfaces

2.1 Recommendation Systems in E-Commerce Practice

The role of recommendation systems, also known as recommender systems or platforms, is to offer suggestions of objects that best suit a given user [1]. Today, these systems are widely utilized in various aspects of life and business, mainly to support in decision-making. Examples of their applications include providing music recommendations, suggesting movies to watch, recommending books to read, facilitating product purchases, and offering diverse content suggestions available on the Web.

One of the most prevalent methods for personalizing content in e-commerce is through product recommendations. The desire to present customers with a list of recommended products can be considered an inherent feature of any form of remote sales. The initial concepts for automating the generation of product recommendations in e-commerce emerged toward the end of the 20th century [2].

Undoubtedly, the foundation of any recommendation system lies in data, which can be explicit, implicit, descriptive, or contextual in nature. Extracting *explicit data*, such as customer reviews of products, necessitates user engagement, where they share their opinions, experiences, or ratings. When users are reluctant to participate in such activities, collecting explicit data can become challenging and may require a substantial amount of time to acquire a learning set.

Implicit data, encompassing on-site activities like page visits, product interactions, searches, and clicks, as well as off-site activities including interactions with emails, ads, mobile applications, and push notifications, is comparatively easier to collect. This data can be automatically gathered based on the interactions between the user and the system. However, a potential challenge in this case is information overload, where a plethora of data is generated, only a portion of which is relevant for analyzing user preferences. Additionally, automatic data collection

A. Wasilewski, *Multi-variant User Interfaces in E-commerce*, Progress in IS, https://doi.org/10.1007/978-3-031-67758-8_2

can be subject to restrictions related to the use of cookies and compliance with GDPR regulations.

Descriptive data, which includes attributes like titles, categories, prices, and textual descriptions, may appear to be the simplest to collect, as they provide a structured representation of a product or service's characteristics. However, it is essential to acknowledge that descriptions from various sources may have diverse structures, vocabularies, and employ synonyms, rendering it challenging to process such data automatically.

Contextual data, encompassing factors such as the device used, location, browser, and session time, can be automatically collected, but the quality of the obtained information is not always of high fidelity. Users' devices and software can either block or falsify the collected information, for example, if a user is using a VPN.

Data collection marks the initial step in developing recommendations, and the subsequent step is to process this data. Among the most widely adopted recommendation approaches are [3, 4]: collaborative filtering, content-based filtering, and hybrid approaches that combine different techniques (Fig. 2.1).

Content-based filtering (CBF) takes into account the characteristics or features of the items being recommended. This approach emphasizes factors that can include textual information like item descriptions, genre, keywords, or metadata, such as director, actors, or release year, especially in the case of movies.

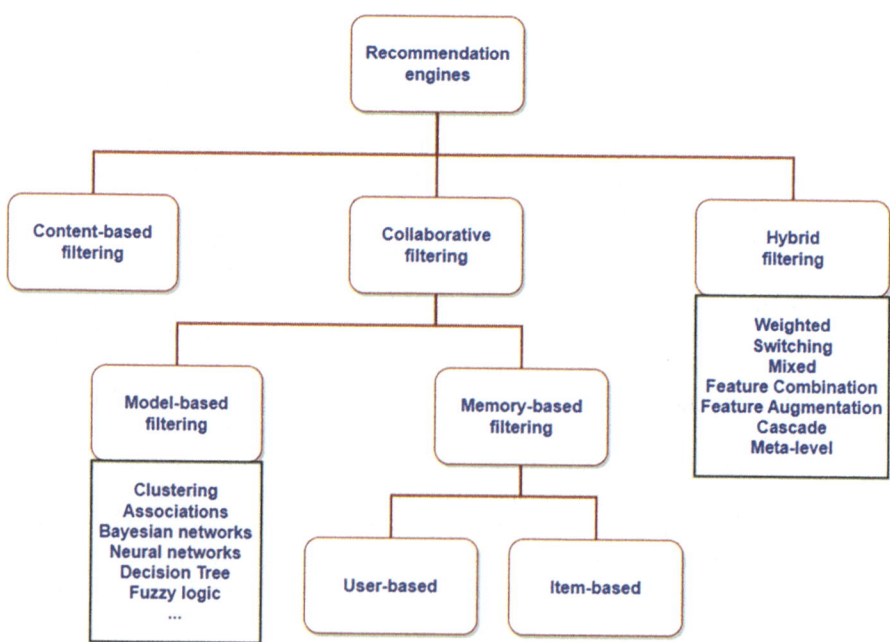

Fig. 2.1 Recommendation techniques [5, 6]

The advantages of this approach are [6]:

- *Independence from user activities*: New items can be recommended even when user ratings or other user-side actions are absent. This is particularly valuable in cases where different users do not share the same items but where items are similar in terms of their intrinsic characteristics.
- *Adaptability*: It can be quickly adjusted as user preferences evolve over time.
- *Transparency*: Recommendations are based on explicit item characteristics, rendering the reasoning behind the recommendations transparent and understandable to users.
- *Quick start*: CBF offers possible recommendations for new users with no history of interactions, as it relies on item characteristics rather than user behavior. This mitigates the "cold start" problem typically associated with user-based recommendation systems.
- *Privacy friendliness*: Users can receive recommendations without the need to share their profile, enhancing the security of personal data [7].

The disadvantages of the CBF include [8]:

- *Dependency on items' metadata*: CBF relies on rich and accurate descriptions of items, and its effectiveness depends on the availability of descriptive data.
- *Computational complexity*: As the number of items and users increases, the computational complexity of CBF can escalate, limiting its scalability for very large datasets.
- *"Cold start" problem*: While CBF does not typically suffer from a "cold start" problem for users, it may encounter such a problem for new items lacking established content features.
- *Risk of overspecialization*: Recommendations provided by CBF may closely resemble items already present in users' profiles, [9], and the model has limited ability to introduce users to new interests.

An example of the practical application of CBF in e-commerce is the *Similar products* section found in e-shops (Fig. 2.2).

SIMILAR PRODUCTS

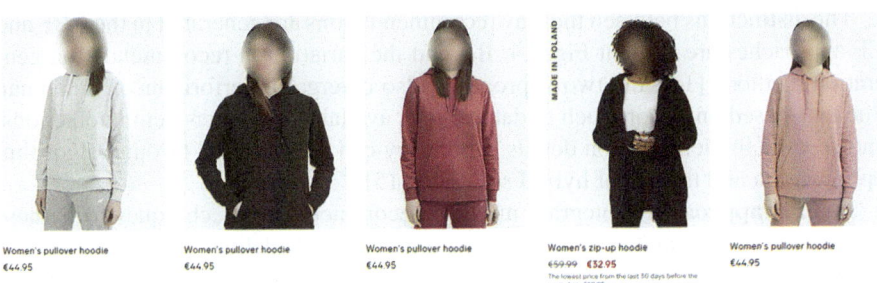

Women's pullover hoodie
€44.95

Women's pullover hoodie
€44.95

Women's pullover hoodie
€44.95

Women's zip-up hoodie
€59.99 €32.95
The lowest price from the last 30 days before the promotion: €59.99

Women's pullover hoodie
€44.95

Fig. 2.2 An example of *Similar products* recommendation (https://4fstore.com/)

Collaborative filtering (CF) operates under the assumption that if two users (user-based collaborative filtering) or two items (element-based collaborative filtering) exhibit similar characteristics, preferences, or past behaviors, their future similarities are likely to align as well [10].

This approach can be further enhanced through the application of machine learning (ML) techniques, including deep learning [11] and clustering methods. These methods are sometimes employed for customer segmentation in e-commerce [12]. This second approach facilitates the grouping of customers based on their similarities and differences, serving as a foundation for the creation of multivariant e-commerce user interfaces.

The main advantages of CF are:

- *Independence from item attributes*: CF does not necessitate in-depth knowledge of item attributes for recommendations. It relies on user behavior and preferences, enabling a wide range of practical applications.
- *Ability to discover relationships*: CF can identify and suggest items based on the preferences of similar users, even if those items fall outside the user's known preferences or familiar item characteristics.

Key CF limitations include:

- *Cold start problem*: CF requires a solid understanding of user relationships and a basic knowledge of items to operate effectively. As a result, its applicability for recommendations to new users is limited.
- *Data sparsity problem*: Generating high-quality recommendations can be challenging when the available information is sparse or limited [13].
- *Scalability*: The computational demands of CF increase linearly with the number of users and items [13], necessitating the use of recommendation techniques that can scale effectively as the dataset's size or complexity grows.

In e-commerce, CF can significantly enhance the customer experience by offering personalized recommendations, which, in turn, boost customer satisfaction and drive sales. It allows users to discover new products that may align with their interests, even if they are not actively seeking them. A practical example of CF's application in an online shop can be observed in the *Other Customers Also Buy* section (Fig. 2.3).

The distinctions between the way recommendations are generated in the CBF and CF approaches are evident Fig. 2.4. Beyond the variation in recommendation generation methods [14], the two approaches also diverge in performance, which can fluctuate based on factors such as dataset size, availability of user–item interactions, and specific implementation details. Efficiency can be enhanced through algorithm optimization and the use of hybrid solutions [15].

Hybrid approaches integrate multiple recommendation techniques to harness their respective strengths, address their limitations, and deliver more accurate and diverse recommendations [16]. These hybrid approaches in recommendation systems offer flexibility and adaptability, making them well-suited to meet the specific requirements and characteristics of the application domain [17]. By employing

OTHER CUSTOMERS ALSO BOUGHT

Fig. 2.3 An example of *Other customer also buy* recommendation (https://4fstore.com/)

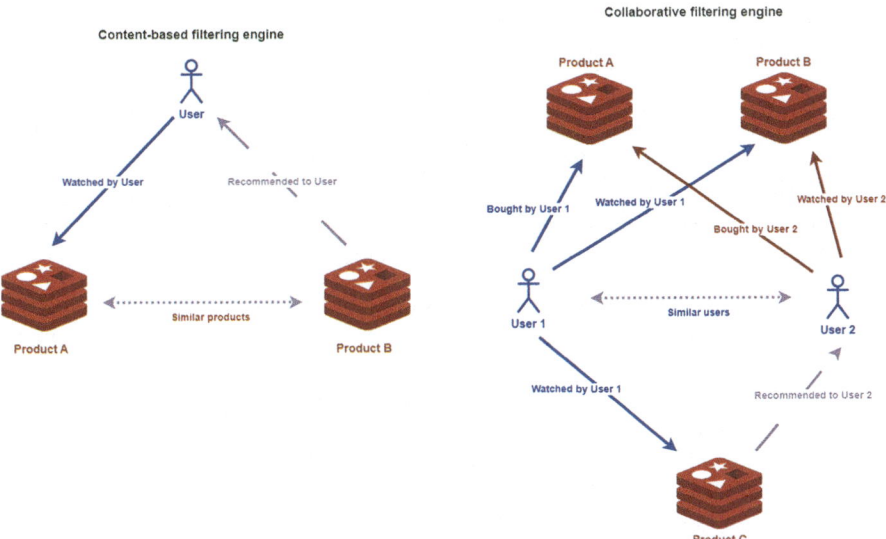

Fig. 2.4 The general difference between CBF and CF approaches

multiple recommendation techniques, hybrid models can mitigate the weaknesses of a single approach [18].

Hybridization techniques include [19]:

- *Weighted hybridization*: Recommendations from different systems are combined by integrating the scores of each technique through a linear formula.
- *Switching hybridization*: Multiple recommendation systems coexist and switch based on accepted heuristics, depending on the conditions under which a recommendation is generated.
- *Cascade hybridization*: Recommendations are generated in successive iterations using different techniques, with the recommendations from the first technique refined by the subsequent one.

SELECTED FOR YOU

Fig. 2.5 An example of *Selected for you* recommendation (https://4fstore.com/)

- *Mixed hybridization*: Recommendations from multiple techniques are presented together, providing a large number of recommendations simultaneously.
- *Feature combination*: Features generated by the first recommendation technique are incorporated as an additional factor in another recommendation technique.
- *Feature augmentation*: The first technique generates a rating or classification of an item, and this information is used in the processing of the next recommendation technique.
- *Meta-level*: The internal model generated by the first recommendation technique is utilized as enriched input for the subsequent technique.

The hybrid approach finds extensive applications in practice [20, 21], and the impacts of such recommendations in an e-shop can be showcased to customers through sections like *Selected for you* (Fig. 2.5) or similar.

In the practice of e-shops, it can be challenging to locate recommendation systems that extend beyond product recommendations. However, the applications of filtering methods transcend conventional online shops and can be adapted to international trade as well [22].

Occasionally, apart from product recommendations, additional methods for personalizing content or UI design can be employed. A common prerequisite for implementing these approaches is the preparation of multiple user interface variants, each of which is supplied with different data or has a distinct structure [23].

2.2   Variants of the E-Commerce User Interface

Today, e-shops offer various approaches to provide more or less personalized user interfaces. These options can be influenced by:

- Customer location
- Special needs of the customer
- Customer segmentation
- Sales model

Serving an interface variant based on the **customer's location** is typically associated with changes in language, currency, or pricing. In such cases, the correct settings for the customer can be automatically retrieved, often through geolocation or IP address, and the customer can be redirected to the appropriate store view or prompted to select the view suitable for their location (Fig. 2.6).

Some shops also permit customers to manually set their location, affecting language and currency preferences (Fig. 2.7), although this is less common.

Sometimes, for legal reasons (e.g., restrictions on the availability of certain products) or marketing considerations (differing pricing policies), customers may not have the option to select their location themselves.

Regional modifications at a general level (Fig. 2.8) primarily pertain to language. Often language specifics (different word lengths to define the same thing) also require structural changes, such as rearranging objects or adjusting layouts. Differences can extend to the menu structure, e.g., varying the basic or special categories (green tags) and colors (purple tags).

At times, distinctions in the interface variants provided to customers from different countries can also extend to the product card (Fig. 2.9). In the example provided, a small section of the product card exhibits five variances. While pricing

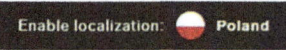

Fig. 2.6 Redirection option (https://4fstore.com/)

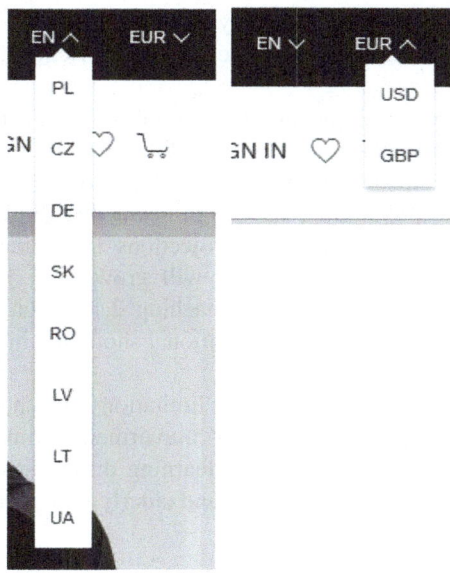

Fig. 2.7 Localization selector (https://4fstore.com/)

Fig. 2.8 Examples of differences at homepage level (https://4fstore.com/)

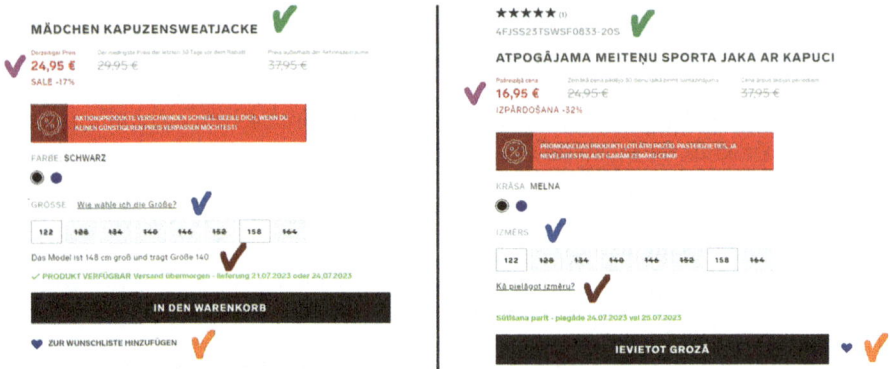

Fig. 2.9 Examples of differences on a product page (https://4fstore.com/)

decisions are typically related to marketing policies, the other modifications pertain to interface layout. These variances may not be substantial, but they have the potential to influence the customer experience and, consequently, the effectiveness of the e-shop.

A complex issue concerns the interface variations that arise from the special needs of customers, particularly in relation to their disability or age. According to the World Health Organization (WHO), approximately 1.3 billion people, which is 16% of the world's population, have significant disabilities [24]. The elderly population is equally substantial, with projections indicating that by 2030, the number of individuals aged 60 and over will grow from 1 billion in 2020 to 1.4 billion, and it will double by 2050, reaching 2.1 billion [25]. Both of these groups may encounter challenges with traditional shopping, making them potential e-commerce customers.

Accessible websites should address their limitations, which may include sensory impairments (hearing and vision), motor impairments (limited use of hands), and cognitive impairments (language and learning disabilities) [26]. Some of the features that can be beneficial for disabled and elderly individuals include [27]:

- Text size enlarge
- Image size zoom
- Text-to-speech for blind and vision-impaired persons

- Speech-enabled searching and selection of items for motor-disabled persons
- Description of image by touch for blind persons
- Closed captioning in videos for hearing-disabled persons
- Contrast versions of the layout
- Alternative pointing devices and keyboards

The World Wide Web Consortium (W3C) has developed the international standard Web Content Accessibility Guidelines (WCAG) [28], which comprises four key principles: (1) perceivable—web page elements and content must be easily perceivable by the senses; (2) operable—users should be able to successfully use all necessary controls, buttons, and other interactive elements, allowing them to navigate the site with ease; (3) understandable—websites should be as user-friendly as possible, making it easy for users to understand and learn how to use them; (4) robust—websites must be capable of supporting usage across a wide range of devices, technologies, and assistive tools, such as screen reading or voice control software. In theory, WCAG recommendations aim to help website developers and designers better meet the needs of all users, including those with disabilities and older users [29]. However, in practice, a 2023 study of one million pages revealed that a staggering 96.3% of them contained WCAG 2 errors. Furthermore, this percentage has only decreased by a mere 1.5% (down from 97.8%) in the past 4 years [30]. This highlights the ongoing need for significant improvements in this area, especially given that 65% of the disabled consumers have abandoned their purchases due to poor accessibility [31].

When analyzing today's e-shops, it is indeed quite uncommon to discover improvements designed to cater to individuals with special needs. Only a limited number of websites offer options to adjust contrast/color schemes or provide tools aimed at enhancing the e-shop's usability for older or disabled individuals. A convenient solution for companies that want to accommodate people with special needs is to use specialized plug-ins that provide accessibility for different types of disabilities. An example of such a solution is the AccessibiltyEnabler developed by HikeOrders (https://hikeorders.com/accessibility/).

As the population of individuals with special needs continues to grow, the awareness of accessibility issues is bound to expand. One potential solution to reach these groups of potential consumers could involve offering them a tailored version of the user interface that is not only more user-friendly but also easier to read and use. These adaptations could significantly benefit people with disabilities or the elderly while still allowing for the possibility of presenting enhanced and impressive user interface variants to other customer groups.

Customer Segmentation The process of dividing a customer base into smaller, more homogenous groups is one of the most critical marketing tools in e-commerce. It offers opportunities for precisely targeting communication and content [32]. Various customer segmentation models exist, including:

- *Demographic segmentation*: This approach is based on demographic information such as age, gender identity, occupation, income level, education, marital status, and more. For example, it could involve segmenting customers into categories like men versus women.

- *Geographic segmentation*: Geographic segmentation divides customers based on their location, such as international versus local customers. This is particularly useful for businesses that serve specific regions or countries.
- *Psychographic segmentation*: This method involves segmenting the market based on personality traits, values, interests, lifestyles, hobbies, and other similar characteristics. For instance, it may consider preferences related to outdoor or indoor activities.
- *Recency, Frequency, and Monetary (RFM) segmentation*: RFM segmentation relies on three key factors: recency of purchases, frequency of purchases, and the monetary value of those purchases. It helps categorize customers into groups like high-value versus low-value customers.
- *Technographic segmentation*: Technographic segmentation is based on the technology customers use, such as mobile devices or desktop users.
- *Behavioral segmentation*: Behavioral segmentation categorizes customers based on their behavior, including buying history, interactions with the website, or engagement on social media channels. An example might be distinguishing between loyal club members and nonmembers.

The decision about dividing customers and selecting the segmentation technique is in the hands of e-commerce proprietors. The examination revealed that they consider behavioral segmentation (38.7%) and mixed segmentation (22.6%) as the most suitable methods [33]. It is significant to note that the development of artificial intelligence applications has led to the extensive use of AI methods like clustering for segmentation [34]. These solutions can surpass basic decision-making rules and identify distinctive and shared traits among customers, even if they are not immediately apparent.

With customers grouped together, messages can be targeted to increase their activity, encourage them to take specific actions, and more. In practice, e-commerce platforms (e.g., Adobe Magento) allow for dynamically displaying content and promotions to specific customers. The dedicated interface variants prepared in this manner typically differ in advertisements, products, or customer promotions presented (e.g., for loyalty program members) but do not affect the overall layout structure. The detailed examination of how customers behave and use the shop, made possible by ML-based data analysis, offers enormous possibilities. By categorizing customers based on their behaviors and the paths they take through the e-shop, groups can be identified that would benefit from using a dedicated interface variant incorporating more comprehensive changes. For instance, if there are customers who read product descriptions carefully, an interface type can emphasize these descriptions. This can also simplify the purchase process for customers and enhance their satisfaction with the online shop.

Sales model can impact the interface version provided, given varying functional requirements. Websites catering to both B2C and B2B sales may exhibit differences

in the design of product cards. The B2C segment of the website might differ from
the B2B section for several reasons:

- The way prices are shown (including taxes or not)
- The availability of products and quantitative limits on goods available for
  purchase
- The approach to the design of the interface (striking visuals vs. simplicity and
  clarity)

Product cards can exhibit substantial variations, particularly in B2B transactions.
They may facilitate bulk purchases, allowing for swift orders or simultaneous
purchase of all product options.

Other examples of features that may affect the UI of B2B systems include:

- Minimum and maximum order units per customer group
- Support for Request for Quote document
- Separate B2B registration form
- Segmentation for dedicated prices and promotions
- Quick order support by importing a file with the ordered products
- Delivery options with dynamically determined cost and route

In practice, it is common to encounter various interface variants for different sales
models or even for different brands of the same manufacturer. These variants,
however, remain static, developed, and implemented during the initial stages of e-
shop development.

The provision of static user interfaces comes with significant drawbacks, primarily related to implementation costs and time constraints. Moreover, it is not
always feasible for the business owner to accurately determine the precise number
of UI variations needed or who should be targeted with them. Therefore, it
is worth exploring the possibility of creating dynamic UI variations that align
with customer classification based on their traits and actions. Addressing this
challenge necessitates a multicomponent solution that encompasses the various
aspects involved in creating and delivering differentiated and dynamic UI variants
in the field of e-commerce. The initial step in developing such a solution should
involve an exploration of established and widely adopted recommender systems.

2.3   Components of a Recommendation System

E-commerce recommendation system architectures typically encompass various
components that collaborate to provide personalized offers to customers [1]. At
a high-level overview, these individual modules are responsible for data acquisition and processing, recommendation generation, and the delivery of these
recommendations to end users. These components are essential not only in exem-
plary recommendation systems applications, such as purchase recommendations
[35], educational hypermedia targeting [36], image recommendation [37], media

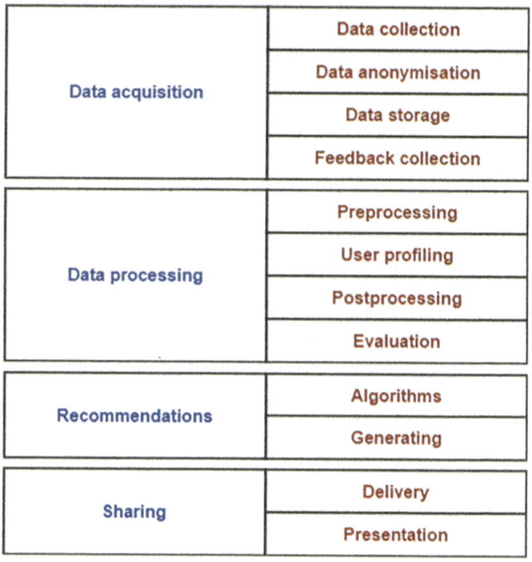

Fig. 2.10 Recommendation
system modules

Data acquisition	Data collection
	Data anonymisation
	Data storage
	Feedback collection
Data processing	Preprocessing
	User profiling
	Postprocessing
	Evaluation
Recommendations	Algorithms
	Generating
Sharing	Delivery
	Presentation

recommendation [38], and suggesting website personalization [39]. The specific architecture details can vary depending on the application's objectives and typically consist of a collection of modules (Fig. 2.10) that are necessary to achieve the stated goals of a particular solution.

Data collection typically serves as the initial step in constructing a recommendation system. This module should facilitate the gathering of data from various sources or systems, as necessitated by the particular requirements of the recommendation system. The data collected may encompass:

- *User data*, including browsing history, purchase history, and demographics
- *Item data*, including descriptions, attributes, features, and categories
- *Contextual data*, including time, location, and devices

Data anonymization is imperative when dealing with personal data subject to privacy regulations and potentially confidential information. Anonymization often involves removing user identification based on obvious information (e.g., email) and substituting it with unique tokens (tokenization).

Data storage typically in the form of a data repository or data warehouse serves as the foundation for housing the collected data. This component is essential for enabling the utilization of the gathered data, ensuring scalability, and facilitating accessibility for subsequent analysis.

Feedback collection plays a crucial role in evaluating the effectiveness of recommendations. Feedback mechanisms can be introduced to enhance the recommendation engine further. Feedback may originate from users directly or result from the calculation and analysis of specific performance indicators. User feedback can take the form of direct ratings or indirect information derived from user behavior, such as order history.

Preprocessing: The process of data preparation involves filtering, cleaning, and transforming data into the appropriate format for subsequent processing. This ensures that the collected data can be effectively processed using selected algorithms. Preprocessing tasks may include removing duplicates, handling missing values, and normalizing the data.

User Profiling is instrumental in creating structured collections of information about users, which are based on their past interests, behavior, and preferences. Various techniques, including machine learning, can be employed to extract relevant characteristics from user data.

Postprocessing, particularly valuable when utilizing a hybrid approach that integrates multiple methods of data analysis, allows for further refinement or optimization of the final results. It helps in eliminating irrelevant or redundant recommendations and prevents the duplication of identical recommendations.

Evaluation, conducted based on feedback, enables the assessment of the recommendation system's performance, effectiveness, and accuracy. This analysis necessitates the use of a set of indicators chosen according to the context of the recommendation engine's use. The results of the evaluation can be used to fine-tune the system or algorithms, thereby enhancing the quality of recommendations.

Recommendation algorithms encompass a set of data processing methods employed to generate recommendations using previously collected information. The selection of algorithms hinges on the purpose of the recommendation, the available data and its quality, and the computing resources at hand. Some algorithm categories, like clustering, may encompass various approaches. During the design phase of a recommendation system, identifying the optimal algorithm (or algorithms) can be a challenging task. Consequently, it is necessary to explore different alternatives before arriving at final decisions. It is also feasible to integrate different algorithms into the recommendation system and subsequently select the most suitable solution based on the context and purpose of the recommendation.

Recommendation generation entails the creation of suggestions based on collected data and the chosen algorithms. The result of this component is a catalog of recommendations that can be presented to the recipient or utilized for further processing.

Delivery of recommendations involves the transfer of the generated recommendations to a system that can present them to the recipient. The method of delivery is dependent on communication channels, such as email or push notifications on mobile devices, and the systems responsible for presenting the results to the user. It is essential to ensure that the data is presented in a suitable format, as the information transmitted from the recommendation system may not be easily understandable to the recipient.

Presentation is the component responsible for delivering user recommendations generated by the recommendation system. It is customized based on several factors, including the device, user interface, user experience, and the user's ability to comprehend the recommendations received.

The components outlined for building the recommender system are crucial, although some aspects, like postprocessing, may be optional. It is worth noting that the crux of recommender systems lies in their ability to enhance the produced outcomes, necessitating iterative repetition of processes and functioning cyclically.

Understanding the motivations behind the implementation of various user interfaces in e-commerce and having knowledge of typical recommendation systems are fundamental to developing a platform concept that would facilitate the delivery of diverse user interfaces in e-commerce solutions. Such a proposal should consider best practices from existing applications but also must align with the objectives and business context.

2.4 An Architecture of Multivariant UI Platform

The implementation of personalized user interface designs in e-commerce can be driven by various needs and objectives. The complexity of the resulting solutions is, in turn, a consequence of the assumptions and expectations that underlie them. The primary method for personalizing the e-commerce interface is to provide different contents, including customized prices, products, and advertising. This approach is relatively straightforward, as it does not necessitate customization of the website layout. The implementation process involves the integration of the back-end system [42], for instance, by retrieving prices based on partner groups from the Enterprise Resource Planning (ERP) system or by configuring and installing one of the available product recommendation engines. This aspect of user interface variability will not be delved into further, as it is commonly and widely used in practice. Another way to personalize the interface is to offer different variants of the user interface. This is a more complex approach that requires a detailed analysis of needs and constraints.

E-commerce interface layouts can be categorized into two variants: static, which is designed and implemented by developers, and dynamic, which can be modified through a configurator accessible to the e-commerce administrator. The easiest approach is to offer static layout variants for customer groups defined by straightforward rules (e.g., logged-in versus not logged-in, loyalty club members versus nonmembers, etc.). Dynamic interface variants require a complex solution that provides high flexibility and applies to multiple business scenarios:

- **Rule-based multivariant interface**: This serves interface variants to groups built on decision rules, with the option to test changes to the user interface (A/B testing).
- **Rule-based multivariant interface with behavioral analysis**: This involves additional collection of behavioral data to better analyze the impact of the dedicated interface.
- **ML-based multivariant interface**: This serves interface variants with clusters built using ML-based algorithms.

- **Self-adaptive multivariant interface**: This offers automatic adaptation of interface options to user behavior in groups facilitated by AI-driven recommendation algorithms.

Each case requires a distinct set of components to provide the desired functionality. However, when considering a platform that fully supports the delivery of multivariant UI, it is important to consider the requirements that arise from each business case while allowing for all potential objectives to be achieved.

Rule-Based Multivariant Interface

The objective of managing interface variants for decision rule-based groups is to empower e-commerce administrators to modify layouts without developer intervention. Therefore, the solution's architecture is relatively straightforward, with only a few components (Fig. 2.11).

To create user interface variants, the *Interface Variant Planner* module is used, which facilitates the design of potential modifications forming user interface variants. The base UI can be used to develop alterations by selecting editable areas within the website. Ideas that were discarded during the initial interface design phase can serve as sources of inspiration for modifications. For instance, consider

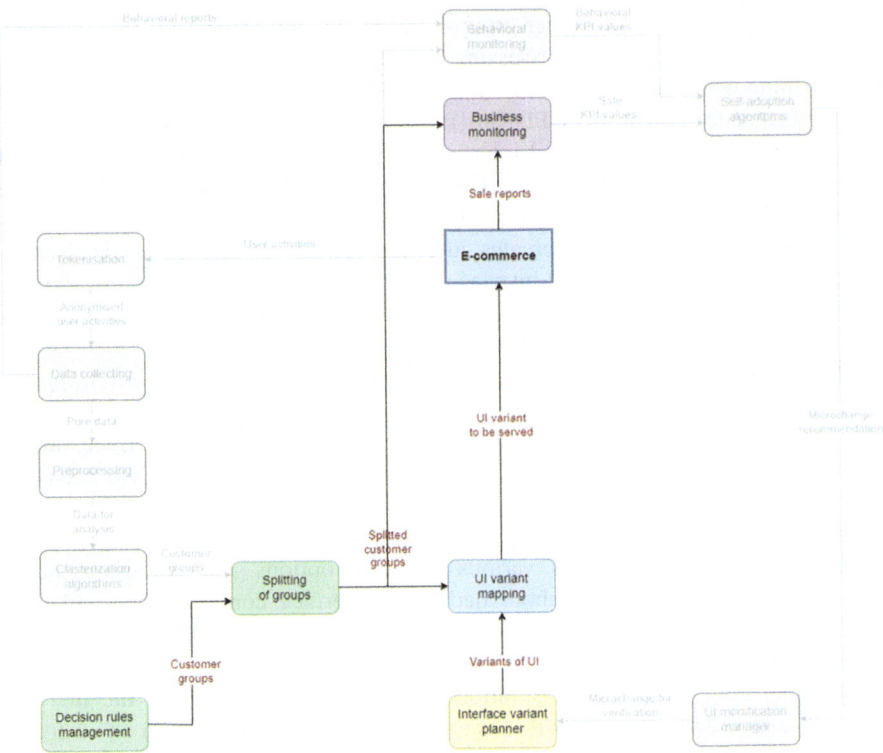

Fig. 2.11 A platform architecture for the rule-based multivariant interface

the search bar, a common feature in online stores. It can be collapsed or expanded, positioned on the left or right side of the layout, marked with an icon or text, and its dimensions can vary. There are numerous possibilities to consider when preparing modification options, and all concepts can be used in different UI variants. The final decision on implementing a particular UI variant rests with UX specialists or, for more advanced solutions, artificial intelligence. Within this component, prospective modifications can be combined to create interface variations, representing the outcome of this phase of the solution.

The next section pertains to the *Decision Rules Management* component of an e-commerce platform, which is responsible for categorizing customers. Criteria commonly employed in e-commerce segmentation, such as the frequency and value of purchases, location, and participation in a loyalty scheme, can be used to group customers. Additionally, a classification system can differentiate between new and returning customers by utilizing stored cookies. The outcome of this system is sets of customers who receive dedicated communication, which may include a user interface tailored to their specific needs. It is imperative to employ the correct method of customer identification to ensure privacy, for example, by avoiding the use of email addresses. This is especially critical in a more complex architecture where the collection of user behavior data plays a pivotal role in the multivariant UI platform.

In specific scenarios, customer groups may be divided using *Split Management*. This division is crucial when assessing the effectiveness of an interface variant. To determine whether the prepared interface variant positively impacts the selected key performance indicators (KPIs [40]), a portion of the customers in the same group should receive a dedicated interface variant, while the other portion should receive a default interface variant. Comparing the KPIs will help identify the superior interface variant to be served to the analyzed group of customers. As mentioned earlier, this module creates customer collections divided into a research subgroup (receiving a dedicated interface) and a comparison subgroup (receiving a standard/default interface). If testing the effectiveness of a user interface variant is unnecessary, this component can be excluded.

The next component within the platform's architecture is *UI Variant Mapping*, where interface variants are linked to client groups for which they are intended. Through this connection, e-commerce users will benefit from an interface variant designed specifically for them by UX experts, based on an analysis of their behavior, preferences, and choices.

E-commerce refers to the e-shop platform used by customers seeking to make purchases. The website offers various interface options and, therefore, needs to accommodate information about dedicated interface configurations and their assignment to customer groups. Additionally, it should have the technical capability to display any user interface modifications that have been made.

The final component of the solution's architecture that supports the first use case is *Business Monitoring*. Its role is to gather essential information on customer purchases, organized by specified groups or subgroups, to calculate sales-oriented KPI values. The e-shop serves as a source of business data as it collects and stores detailed information on placed orders. Analysis of the resulting KPI values enables confirmation of whether the implemented interface variants have increased sales in the online store and should be retained or, conversely, have had a detrimental impact and should be removed.

The presented architecture offers a viable solution for creating a dedicated interface that caters to select groups of customers based on specific decision rules. Additionally, by incorporating the ability to split groups, it facilitates the implementation of A/B testing, one of the most widely used methods for verifying changes made to an e-commerce interface [41]. Due to its simplicity, this solution can be applied to nearly all e-commerce websites, even those with limited data that may be insufficient for advanced computing algorithms.

Rule-Based Multivariant Interface with Behavioral Analysis

The next step in extending the architecture of the platform to support a multivariant user interface is to incorporate components responsible for collecting and basic processing of information related to customer activity on the e-commerce site (Fig. 2.12). Implementing such a solution allows for the expansion of conventional

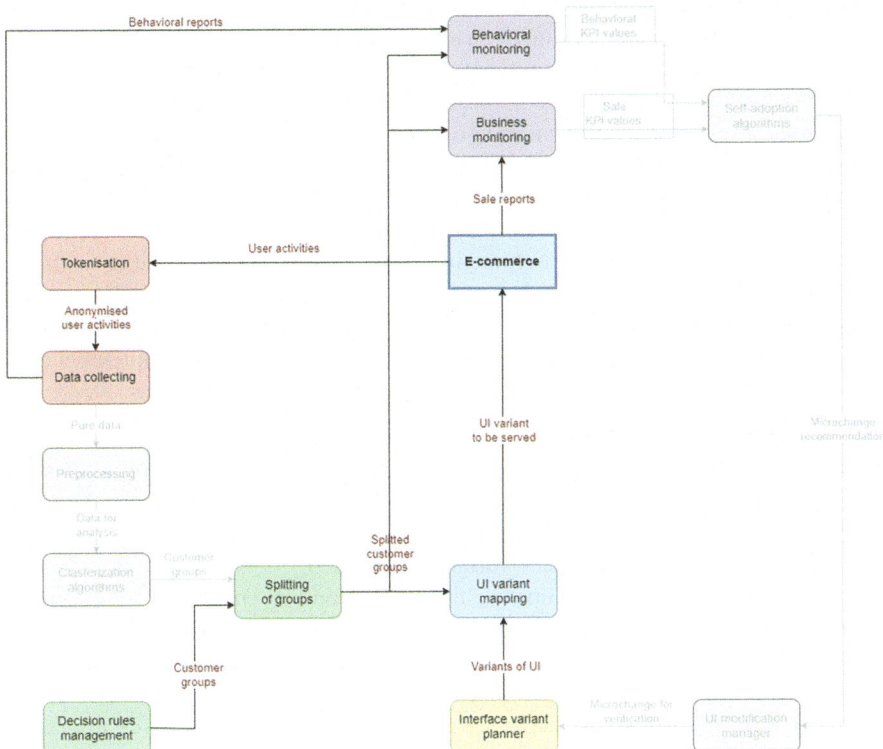

Fig. 2.12 A platform architecture for the rule-based multivariant interface with behavioral analysis

performance analysis beyond sales results alone, incorporating metrics such as Average Basket Value (ABV), Average Basket Size (ABS), and Average Selling Price (ASP). This analysis considers not only customers' purchases but also their use of the e-shop and their actual journey, as well as their preference for supplementary content such as blogs, forums, and additional descriptions.

The initial supplementary element is *Tokenization*, which is responsible for ensuring privacy. This is especially significant in countries where the privacy of Internet store customers is legally regulated. Before storing the observed user actions in the e-shop, a procedure should be implemented to delete or anonymize personal data. The observations can be linked to individual tokens assigned to e-shop customers but without the possibility of decoding or linking to a specific person.

The *Data Collecting* module is responsible for recording customer behavior information, along with contextual details (e.g., date and time of activity). Various customer actions can be documented, including visiting the site, selecting a product (viewing a product card), choosing product features (e.g., color or size), accessing additional information in a pop-up window, adding a product to the basket, using the search engine, and more. Actions are recorded during the user's session, enabling a comprehensive map of their entire e-commerce journey.

The final component of the discussed solution is *Behavioral Monitoring*, which involves the statistical examination of customer conduct, encompassing their actions and sequences of actions. The outcomes of this analysis can be represented in the form of statistical reports, demonstrating the frequency or number of page visits, product selections, color, or size preferences. Performance indicators can also assess nonpurchase behavior, such as whether the observed behavior aligns with the intended customer journey, which includes a set of pages on the e-commerce platform that customers should visit to achieve potential business goals.

This solution, based on the architecture discussed, shares similarities with the capabilities offered by the previous version but extends it to include data on nonpurchase-related customer behavior. Such information can be used to examine the e-shop's day-to-day operations and provides a foundation for the implementation of advanced analytical methods.

ML-Based Multivariant Interface
Gathering comprehensive data on customer behavior within the e-shop allows for analysis using machine learning methods. Extending the architecture with additional components is necessary to classify customers, divide them into groups, and then serve them with dedicated interface variants (Fig. 2.13).

In contrast to previous solutions, the allocation of customers into groups (segments) will not be based on formal decision rules or arbitrary decisions made by the system administrator, but instead on the implementation of specific clustering techniques.

The first component to be added is *Preprocessing*, which involves preparing collected data for analysis by structuring and cleansing it. This module ensures that irrelevant data is filtered out, resulting in a more informative dataset. The main goals of this stage include filling in missing data, summarizing information, and assigning

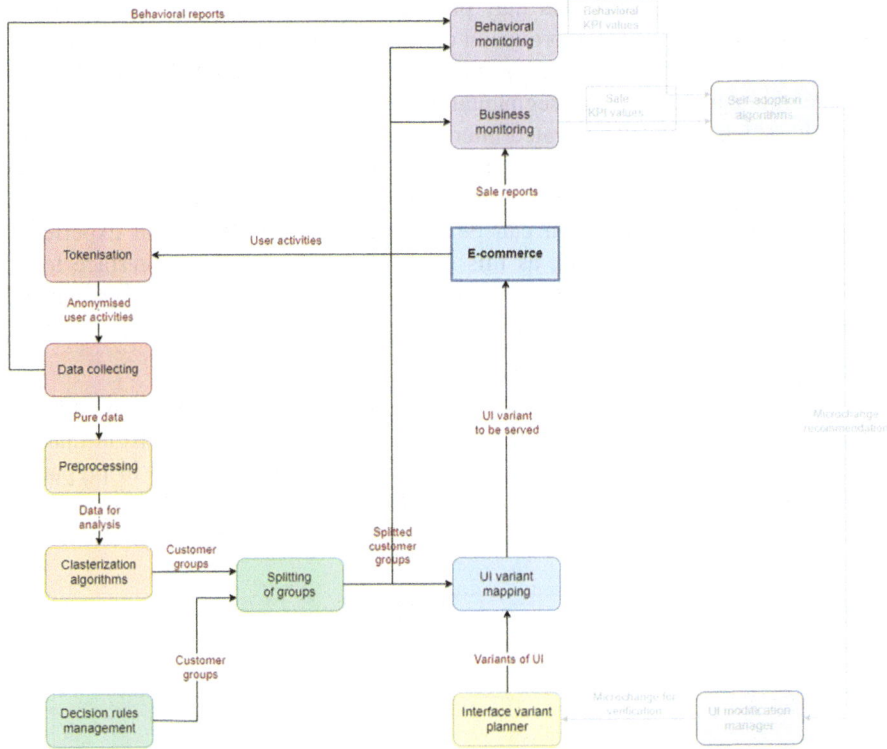

Fig. 2.13 A platform architecture for the AI-based multivariant interface

data to categories. In e-commerce, combining information about user behavior (e.g., accessing a specific product card) with static information (e.g., definitions and product descriptions) can be achieved through preprocessing.

The second component, *Clasterization algorithms*, comprises algorithms that implement selected clustering approaches using ML. Clustering approaches can be classified into distinct categories based on their fundamental principles and characteristics. One primary challenge is the selection of algorithms implemented in this module. Utilizing all available methods, as well as their extensions and adaptations, would be impractical. Therefore, it is crucial to carefully choose appropriate algorithms, considering both traditional measures of clustering quality and the specific requirements of the context, which involve preparing specific groups of e-commerce customers who will be using the dedicated user interface.

While the system's range of applications based on this architecture is similar to previous solutions, the distinction lies in the preparation of customer groups. Segmentation is based on behavioral data encompassing all customer activities on the e-shop. Furthermore, interface variants are customized to the characteristics of the generated customer clusters. Advanced data analysis identifies the distinguishing features of customer groups that may be imperceptible to humans, thus enabling customer segmentation for the provision of bespoke interfaces.

It is worth noting that clustering customers based on their behavior does not preclude the parallel use of segmentation based on decision rules. A hybrid approach, which provides dedicated interface variants to different customer groups generated using machine learning and suitable interface variants to new customers who have not been subject to clustering, can be considered a special case of such an approach. A hybrid solution can be advantageous for stores that experience infrequent returning customers, as this approach allows the user interface to be personalized to meet the unique needs of different groups of customers (known and unknown), maximizing their satisfaction.

Self-adaptive Multivariant Interface

The most complex part of the system architecture is the one that allows for the self-adaptation of user interface variants (Fig. 2.14). In this case, decision-making algorithms supersede the prior expertise of UX specialists involved in preparing interface versions. The algorithms aim to iteratively alter the interface versions

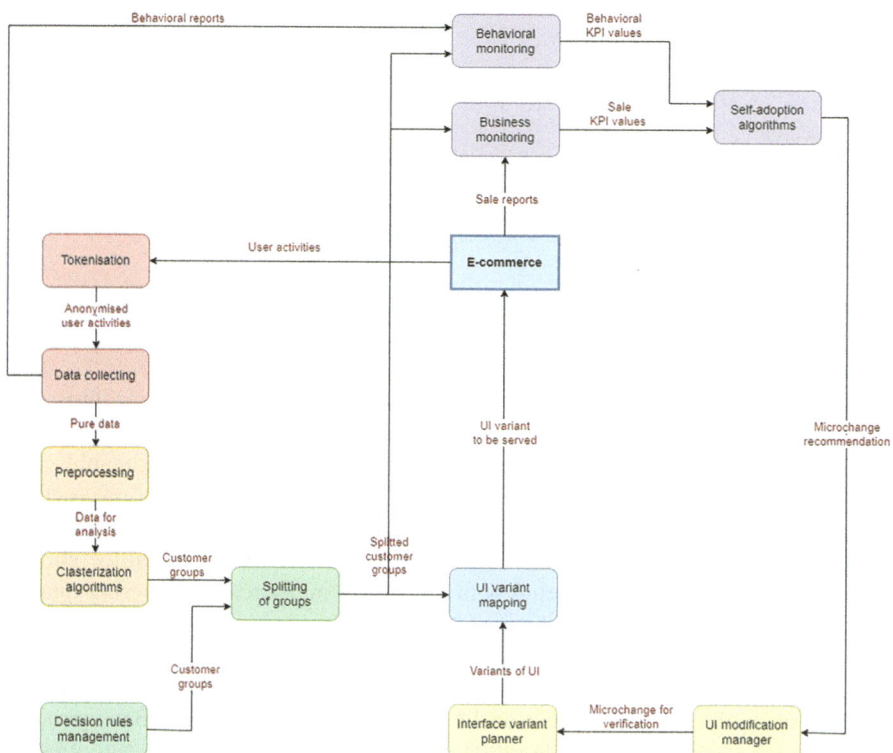

Fig. 2.14 A platform architecture for the self-adaptive multivariant interface

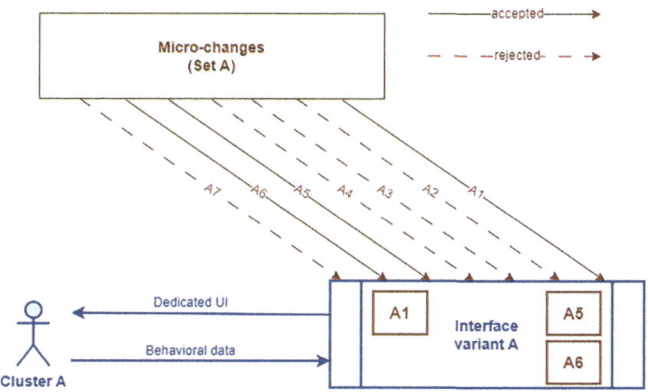

Fig. 2.15 The adaptation of microchanges to the UI variant

catering to various customer groups and attain combinations that optimize the e-shop's efficiency.

The concept of self-adaptation relies on the idea of "microchange," defined as a modification to an interface variant aimed at improving UI efficiency. A microchange may involve tweaking a single interface element (such as collapsing or expanding filters, resizing the search engine, moving the add-to-cart button, etc.) or a group of related UI elements. Microchanges accepted during the assessment process result in a modification of the user interface variant, while rejected changes are disregarded (Fig. 2.15).

The first extension of this version of the system architecture involves the incorporation of the *UI Modification Manager* component. This feature includes the management of microchanges that can be tailored to interface variations catering to particular user clusters. While the list of microchanges may remain consistent across all interface types, certain customer groups may have access to different modifications (Fig. 2.16).

The second extension features *Self-adaptation Algorithms* that employ specific rules to enable the system to determine the approval or rejection of a microchange. Sales or behavioral indicators can be utilized to evaluate such a modification. The decision regarding a microchange can be entirely automatic, semiautomatic (decisions are made that the system can make with acceptable certainty, while doubts are resolved by a human expert), or manual (the system calculates a set of indicator values and a recommendation, but a human expert makes the final decision).

Discussed approaches to e-commerce platform architecture that aim to serve multivariant user interfaces are based on incremental development. This eliminates the need for costly investment in complex solutions at the outset and enables systematic implementation of additional components when they are required.

To validate the proposed architecture and approach in practice, a platform was developed to validate the solution concept and design. The key modules were

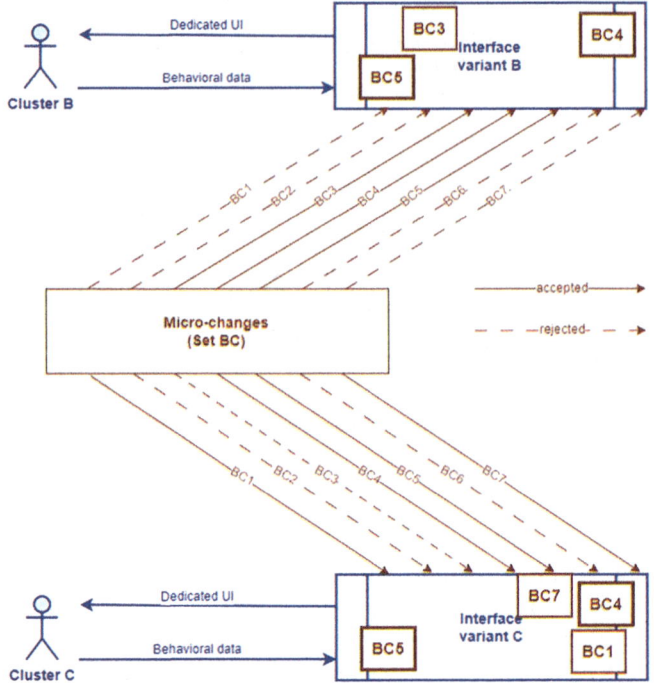

Fig. 2.16 The adaptation of multiple UI variants from a single set of microchanges

integrated, the correctness of the business requirements and the technical and technological assumptions were experimentally tested, and the necessary adjustments
and additions were made. The final version of the platform architecture is the result
of iterative research and development, and the insights gained from the results
identify the strengths and limitations of the proposed concept.

References

1. Ricci F, Rokach L, Shapira B (2022) Recommender systems handbook. Springer, New York
2. Schafer B, Konstan J, Riedl J (1999) Recommender systems in E-commerce. In: 1st ACM
   conference on electronic commerce. https://doi.org/10.1145/336992.337035
3. Jafarkarimi H, Sim ATH, Saadatdoost R (2012) A naive recommendation model for large
   databases. Int J Inf Educ Technol. https://doi.org/10.7763/IJIET.2012.V2.113
4. Yan K (2023) A review of techniques used in e-commerce recommendation system. In:
   Proceedings of the 3rd international conference on signal processing and machine learning.
   https://doi.org/10.54254/2755-2721/4/2023364
5. Roy D, Dutta M (2022) A systematic review and research perspective on recommender
   systems. J Big Data. https://doi.org/10.1186/s40537-022-00592-5
6. Isinkaye FO, Folajimi YO, Ojokoh BA (2015) Recommendation systems: principles, methods
   and evaluation. Egypt Inform J. https://doi.org/10.1016/j.eij.2015.06.005

7. Lam SKT, Frankowski D, Riedl J (2006) Do you trust your recommendations? An exploration of security and privacy issues in recommender systems. In: Emerging trends in information and communication security. Springer, Berlin, Heidelberg

8. Adomavicius G, Tuzhilin A (2005) Toward the next generation of recommender system. A survey of the state-of-the-art and possible extensions. IEEE Trans Knowl Data Eng. https://doi.org/10.1109/TKDE.2005.99

9. Zhang T, Vijay SI (2002) Recommender systems using linear classifiers. J Mach Learn Res. https://doi.org/10.1162/153244302760200641

10. Guo G (2022) Application of E-commerce personalized recommendation algorithm based on collaborative filtering. In: Cyber security intelligence and analytics. https://doi.org/10.1007/978-3-030-97874-7_140

11. Laksana MO, Maulani IE, Munawaroh S (2023) Development of E-commerce website recommender system using collaborative filtering and deep learning techniques. Dev J Comm Service. https://doi.org/10.36418/devotion.v4i2.417

12. Ciguene R, Marron B (2021) Clustering and e-commerce: towards a crossroads in a particular context: categorization of amazon products problematic in intermediation agencies a new context for the use of clustering in e-commerce. In: ICEEG '21: proceedings of the 5th international conference on e-commerce, e-business and e-government. https://doi.org/10.1145/3466029.3466697

13. Park DH, Kim HK, Choi IY, Kim JK (2012) A literature review and classification of recommender systems research. Expert Syst Appl. https://doi.org/10.1016/j.eswa.2012.02.038

14. Liao M, Sundar SS (2020) When E-commerce personalization systems show and tell: investigating the relative persuasive appeal of content-based versus collaborative filtering. J Advert. https://doi.org/10.1080/00913367.2021.1887013

15. Vishwas N, Deb T, Saha A, Kumari L (2020) Implementation of collaborative filtering for product recommendation in e-commerce to enhance scalability and performance. In: Performance management of integrated systems and its applications in software engineering. https://doi.org/10.1007/978-981-13-8253-6_6

16. Cai X, Hu Z, Zhao P, Zhang W, Chen J (2020) A hybrid recommendation system with many-objective evolutionary algorithm. Expert Syst Appl. https://doi.org/10.1016/j.eswa.2020.113648

17. Mandalapu SR, Narayanan B, Putheti S (2023) A hybrid collaborative filtering mechanism for product recommendation system. Multimed Tools Appl. https://doi.org/10.1007/s11042-023-16056-8

18. Abbas AR, Ashor S (2017) Designing personalized recommendation in e-commerce site based on content-based and collaborative filtering. J. Al-Nahrain Univ. https://doi.org/10.22401/JNUS.20.2.19

19. Burke R (2002) Hybrid recommender systems: survey and experiments. UUser Model User-. https://doi.org/10.1023/A:1021240730564

20. Chornous G, Nikolskyi I, Wyszynski M, Kharlamova G, Stolarczyk P (2021) A hybrid user-item-based collaborative filtering model for e-commerce recommendations. J Int Stud. https://doi.org/10.14254/2071-8330.2021/14-4/11

21. Balmadres JAT, Bartolome K, Bunyi RGB, Jacobo JRB, Lalata J-A, Lagman A, Fernando-Raguro MC (2023) Development of hybrid personalized e-commerce using collaborative filtering and content-based filtering for South Cartel clothing company. In: Intelligent sustainable systems, selected papers of WorldS4. https://doi.org/10.1007/978-981-19-7660-5_8

22. Wu X, Wu Z (2023) Application of big data search based on collaborative filtering algorithm in cross-border e-commerce product recommendation. Soft Comput. https://doi.org/10.1007/s00500-023-08643-6

23. Wasilewski A (2024) Functional framework for multivariant e-commerce user interfaces. J Theor Appl Electron Commer Res. https://doi.org/10.3390/jtaer19010022

24. WHO (2022) Global report on health equity for persons with disabilities. https://www.who.int/publications/i/item/9789240063600. Cited 15 Jul 2023

25. WHO (2022) Ageing and health. https://www.who.int/news-room/fact-sheets/detail/ageing-and-health. Cited 15 Jul 2023
26. Sohaib O, Kang K (2017) E-commerce web accessibility for people with disabilities. In: Complexity in information systems development. https://doi.org/10.1007/978-3-319-52593-8_6
27. Hussain MA, Ahsan K, Iqbal S, Al Hassan AN, Sarim M (2016) Assisting disabled persons in online shopping: a knowledge- based process model. J Basic Appl Sci. https://doi.org/10.6000/1927-5129.2016.12.04
28. W3C (2022) Web content accessibility guidelines (WCAG) 2.1. https://www.w3.org/TR/WCAG21/. Cited 15 Jul 2023
29. Filipe F, Pires IM, Gouveia AJ (2022) Why web accessibility is important for your institution. Proc. Comput Sci. https://doi.org/10.1016/j.procs.2023.01.259
30. WebAIM (2023) The WebAIM million. https://webaim.org/projects/million/. Cited 15 Jul 2023
31. Ormesher E (2022) E-commerce is failing disabled people, say experts. https://www.thedrum.com/news/2022/09/26/e-commerce-failing-disabled-people-say-experts. Cited 15 Jul 2023
32. Chen Y, Kuo M, Wu S, Tang K (2009) Discovering recency, frequency, and monetary (RFM) sequential patterns from customer's purchasing data. Electron Commer Res Appl. https://doi.org/10.1016/j.elerap.2009.03.002
33. Zaric S (2021) 9 customer segmentation tips to personalize ecommerce marketing and drive more sales. databox. https://databox.com/ecommerce-customer-segmentation-tips. Cited 15 Jul 2023
34. Gomes MA, Meisen T (2023) A review on customer segmentation methods for personalized customer targeting in e-commerce use cases. Inf Syst e-Bus Manag. https://doi.org/10.1007/s10257-023-00640-4
35. Oldridge E (2022) Recommender systems, not just recommender models. NVIDIA Merlin. https://medium.com/nvidia-merlin/recommender-systems-not-just-recommender-models-485c161c755e. Cited 15 Jul 2023
36. Kristofic A, Bielikova M (2005) Improving adaptation in web-based educational hypermedia by means of knowledge discovery. In: HYPERTEXT 2005, proceedings of the 16th ACM conference on hypertext and hypermedia, September 6–9, 2005, Salzburg, Austria. https://doi.org/10.1145/1083356.1083392
37. Melo EV (2018) Improving collaborative filtering-based image recommendation through use of eye gaze tracking. Information. https://doi.org/10.3390/info9110262
38. Amatriain X, Basilico J (2013) System architectures for personalization and recommendation. Netflix TechBlog. https://netflixtechblog.com/system-architectures-for-personalization-and-recommendation-e081aa94b5d8. Cited 15 Jul 2023
39. Baraglia R, Silvestri F (2007) Dynamic personalization of Web sites without user intervention. Commun ACM. https://doi.org/10.1145/1216016.1216022
40. de Ven M, Machado PL, Athanaspopoulou A, Aysolmaz B, Turetken O (2023) Key performance indicators for business models: a systematic review and catalog. Inf Syst e-Bus Manag. https://doi.org/10.1007/s10257-023-00650-2
41. Siroker D, Koomen P (2015) A/B testing: the most powerful way to turn clicks into customers. Wiley, London
42. Wasilewski A (2019) Integration challenges for outsourcing of logistics processes in e-commerce. In: Asian conference on intelligent information and database systems. https://doi.org/10.1007/978-3-030-14132-5_29

Chapter 3
Analysis of Customer Behavior

3.1 Data Collection

To effectively implement recommendation systems, amassing user behavioral data is indispensable, especially for a platform that integrates multiple user interfaces. This necessity remains unchanged whether the platform encompasses various interfaces. Nevertheless, due to evolving legal constraints concerning privacy and the evolving landscape of third-party cookies, the primary source for data acquisition lies within the e-commerce platform itself, which includes the e-shop and supplementary websites such as the blog and brand zone.

Collecting data on orders placed through an e-commerce platform is a straightforward process since they are meticulously recorded and come with statuses that facilitate progress assessment. However, gathering data on customer actions within the e-shop can pose challenges. In such instances, businesses have two options: They can either develop their proprietary tools for tracking customer behavior or opt for existing solutions offered by specialized vendors. Nevertheless, the integration of records from different sources to furnish comprehensive and reliable information can be problematic. This complexity arises from the necessity to use customer identification methods without breaching privacy and compliance regulations governing the collection and processing of personal data.

User Identification
The first step in gathering data on e-commerce customer behavior is to ensure that customers are uniquely identified and labeled in a manner that allows for linking to their actions while adhering to privacy policies. The unique identifier can be randomly generated or may incorporate contextual information, such as the customer's device specifications or email address.

When using data from a customer's device, such as their browser, operating system, or device type, to create a *device-based token*, it is important to consider limitations. These limitations include the possibility of multiple customers using

A. Wasilewski, *Multi-variant User Interfaces in E-commerce*, Progress in IS, https://doi.org/10.1007/978-3-031-67758-8_3

devices with identical characteristics and the potential for a customer to use multiple devices. Before adopting this approach, it is essential to conduct an objective assessment of the risks associated with potential falsification of personalization effects due to problems related to using devices beyond a simple *one-customer, one-device* scheme.

Using a customer's email address to create an *email-based token* resolves the problem of using multiple devices, albeit with certain constraints. Firstly, it is important to note that an email address can be considered as personal data, and its use must comply with the law. In particular, it should not be possible to reverse the tokenization process and retrieve the customer's email address. Additionally, since the email address is usually provided at the end of the purchase process, there may be challenges in linking the customer's actions before they enter their email address.

Due to contextual limitations with device-based and email-based tokens, it is optimal to use randomly generated tokens. To generate such tokens, a well-known method called a UUID can be applied. This method is based on the standard delivered by the Open Software Foundation (OSF) as part of the Distributed Computing Environment (DCE) [1], included in ISO/IEC 9834-8:2014 [2]. UUIDs are usually represented as hexadecimal values, and the commonly used format is 8-4-4-4-12, which gives sequence structure: xxxxxxxx-xxxx-xxxx-xxxx-xxxxxxxxxxxx (e.g., 708d1e23-b84c-4a9e-9350-615c099a2889). Tokens generated in this way practically guarantee 100% uniqueness of the identifier. The quantity of randomly generated version-4 UUIDs required for a 50% likelihood of at least one collision is 2.71 quintillion [3].

When customers visit an e-commerce website, their cookies can be checked to identify them. If no cookie is found, a *user token* will be created and stored as a cookie on the user's computer. It is important to specify a long *expiration tag* for the cookie, such as 180 days. If a cookie is discovered, it should be extended, and the obtained user token can be used as the customer ID for recording actions on the e-shop. It is worth noting that using a cookie with a *session* tag to identify the client should be avoided, as it will expire when the session ends.

When using UUIDs, it is important to remember that some users may access the platform on devices with cookies blocked. Consequently, these individuals will not be correctly identified, and as a result, the mechanisms for analyzing individual customer behavior and decisions will not function as intended.

Furthermore, it is also possible that customers may delete their cookies or not return to the online shop within the validity period. In either case, when they return, they will not be identified, and they will be treated as new customers, resulting in a new customer token being issued.

User Tracking

To ensure accurate tracking of user behavior on an e-commerce platform, it is essential to use the right tools. Instead of creating a bespoke system, which can be costly and time-consuming, you can leverage various free and paid tools that are readily available. A common and effective approach for collecting consumer data is to combine a tag management system with a web analytics system.

A *Tag management system* (TMS) is software designed to manage tracking tags. Tags are small pieces of JavaScript code that capture the activity data of website visitors and are inserted into the website's source code. Utilizing a TMS can greatly simplify the implementation and maintenance of tags, which are used in online content to communicate with applications such as web analytics, personalization, and advertising. Some of the top TMS vendors include Google Tag Manager, Tealium iQ Tag Management, Adobe Experience Platform Launch, Qubit, and Signal Customer Intelligence Platform [7].

Among the available TMS tools on the market, Google Tag Manager (GTM) is widely acknowledged as the leading option, as evidenced by reports from G2 (Fig. 3.1).

Google Tag Manager, launched in 2012, is a free solution provided by Google that enables the collection of various analytical data from different tools. By implementing a single script (in two parts) in the code of a page, it is possible to support diverse mechanisms and configure data collection capabilities. GTM not only allows you to connect scripts from other Google services but also integrates

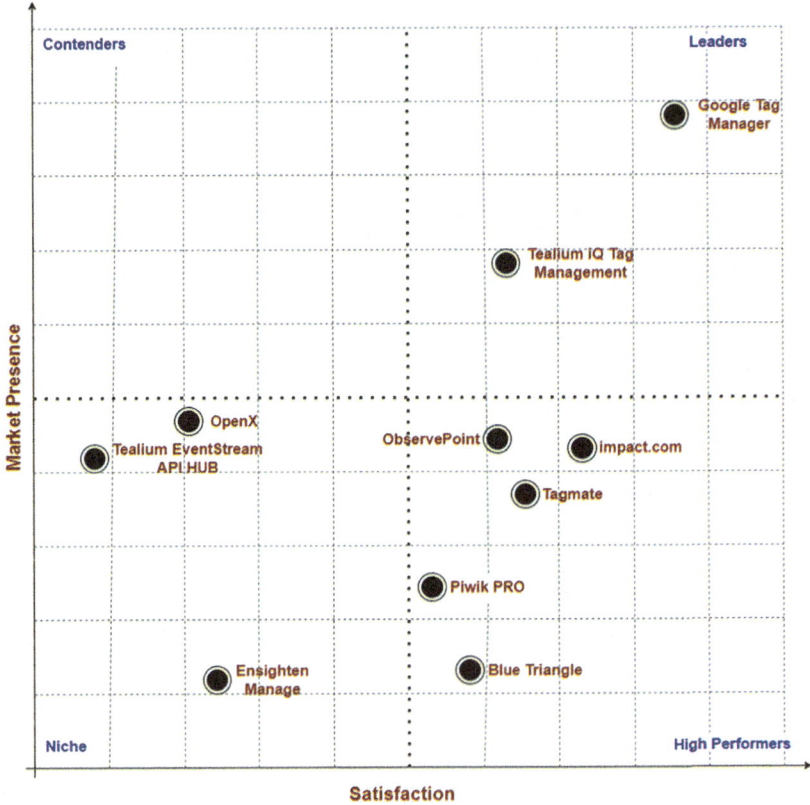

Fig. 3.1 G2 Grid for the top Tag Management Systems products [4]

other external applications such as HotJar, Infinity Tracking, Mouseflow, and custom solutions. GTM empowers marketers and analysts to have a more direct role in the tagging process, making it easier to add, delete, and update tags without the need to directly edit the website code [5]. The activities that can be tracked include clicks on-page elements like links to downloads or external websites, interactions with embedded videos, and social media buttons on a page [6].

Web analytics systems are used to track and analyze user behavior on websites. The technology used is based on cookies or log files, but hybrid solutions are also available. The most popular system in this category is Google Analytics (GA), which holds a market share of more than 50% [8]. Other popular tools are Facebook Pixel, WordPress Jetpack, Yandex.Metrica, Hotjar, MonsterInsights, and Matomo.

Google Analytics, despite its status as the most widely used web analytics solution [9], presents notable limitations when employed to gather user behavior data to provide dedicated UI. The first problem is GDPR compliance, which GA does not provide by default [10]. While it is feasible to adapt GA to meet GDPR requirements, doing so involves intricate steps that typically surpass the competence of the average user of this system [11].

The second issue is the location where GA stores the data, which is in the cloud, leading to data being stored outside the organization's direct control. This situation can be seen as an inconvenience due to the potential collection and storage of sensitive data.

The third issue pertains to GA's delivery model, limiting users' capacity to make extensive customizations to meet specific needs for data collection to support a dedicated UI.

Given these limitations, alternative solutions that are free from the aforementioned drawbacks can be considered, with Matomo (formerly, Piwik) deserving particular attention. Matomo is an open-source analytics platform that places a strong emphasis on user privacy [12]. Its distribution model allows for customization, and data is collected locally in its own database (MySQL). When used in combination with GTM, Matomo enables the collection of various types of information about customer actions in the e-shop, allowing for comprehensive tracking of customer activity while adhering to privacy rules.

An example of the data collected by Matomo is shown in Fig. 3.2. This data ranges from contextual information (device, location, etc.) to detailed insights about the pages visited and decisions made by the customer. Furthermore, it is possible to link the activity history to a specific user by associating the data with a randomly generated UUID (top right).

Another notable way to implement web analytics is through *clickstream* user behavior analysis [13]. This approach is valuable for understanding user engagement, identifying areas for improvement, and optimizing the user experience. It features a collection of detailed information, including page views, clicks, mouse movements, scroll depth, form submissions, and more. The effects of such a solution can be:

• *Path analysis*: This involves examining the sequences of pages users follow, identifying common paths, and pinpointing drop-off points [14].

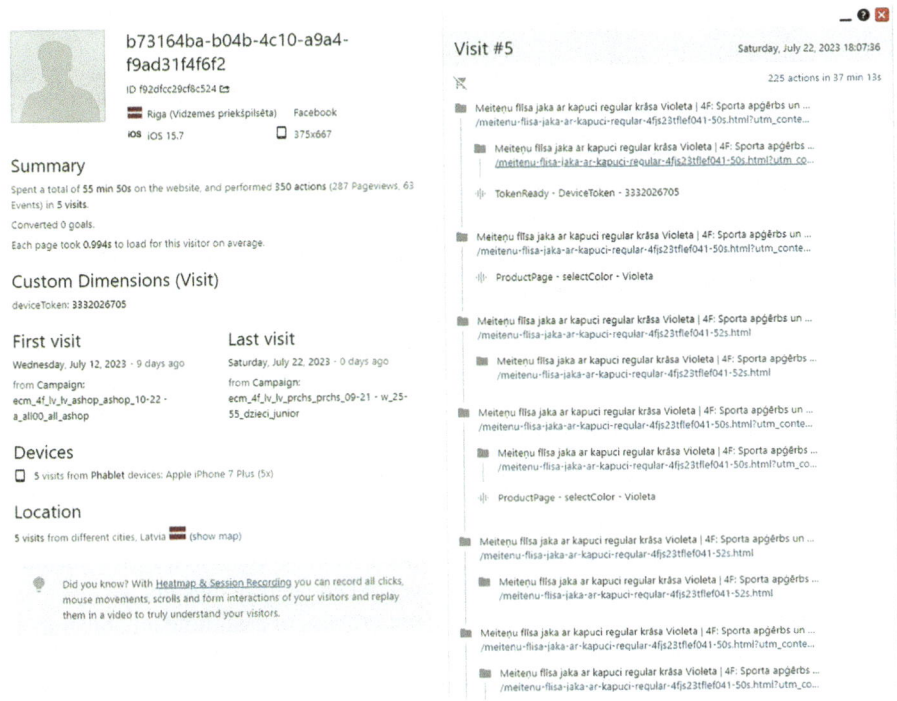

Fig. 3.2 User data collected by Matomo

- *Funnel analysis*: Funnel analysis tracks the step-by-step flow of users through a specific process to identify areas with high abandonment rates [15].
- *Segmentation*: Users can be divided based on various attributes (e.g., new vs. returning users, geographic location), and their behavior can be analyzed separately [16].
- *User flow analysis*: This provides a visualization of the typical paths users take through the site, including entry and exit points [17].

Clickstream user behavior analysis provides valuable insights for website owners, marketers, and UX designers. It aids in understanding how users interact with their online assets and empowers them to make data-driven decisions to improve the user experience and achieve their business goals. This approach requires the collection of user behavior data, which is analogous to the data required to operate a complex system supporting a multivariant user interface.

The process begins with gathering information about user behavior, which is the first step in preparing user groups to be served by a dedicated user interface. The second step involves analysis. In addition to traditional segmentation based on decision rules, methods derived from machine learning can also be employed. One way to create user groups (clusters) using ML is by performing the activities that constitute the cluster analysis process.

3.2 Process of Cluster Analysis

Clusterization (also known as clustering or cluster analysis) is an unsupervised learning method that groups a series of objects, such as customers, in a way that objects within the same group (referred to as a cluster) are more similar, as defined by a specific criterion, to each other than to objects in different groups or clusters. This approach aims to identify inherent structures, patterns, or relationships in the data without relying on any pre-established labels or categories [18]. One of the numerous applications of clustering is in recommender systems, which are used to forecast a user's preferences based on the preferences of other users.

The process of clusterization includes several steps [19]:

1. *Data representation*: This step involves the collection and preprocessing of data to prepare the dataset for further analysis.
2. *Proximity measurement*: Proximity measurement entails using a distance or similarity/dissimilarity metric to determine the degree of similarity between two data points [20].
3. Selection of the *clustering algorithm*: Objects can be classified based on the cluster model, which determines the approach taken to form clusters.
4. *Cluster assignment*: Cluster assignment is the process of assigning an object to a specific cluster based on its similarity to other objects in that cluster.
5. *Evaluation*: This step involves assessing the quality of the clustering results using various criteria, such as within-cluster similarity and between-cluster dissimilarity, to ensure that the created clusters are meaningful and useful for the intended application.

The first step of the process was discussed in the previous section, but subsequent steps need to be clarified.

Proximity Measurement
Within the clustering process, the calculation of *proximity* (similarity/dissimilarity) between two data objects is of paramount importance. In this context, the goal is to quantify the similarities and differences among the characteristics of the analyzed objects, enabling the identification of patterns that facilitate their grouping. It is worth mentioning that factors such as data sparsity, correlations among data features, and feature formats can potentially obscure meaningful relationships during this step [21]. Popular proximity measures widely used in clusterization are:

- *Euclidean distance*
- *Cosine similarity*
- *Jaccard index* (Jaccard similarity)
- *Manhattan distance* (taxicab geometry, city block distance)

Euclidean distance is used in many clustering algorithms including K-means and hierarchical clustering and is calculated (in n-dimensional Euclidean space):

$$d(p, q) = \sqrt{(p_1 + q_1)^2 + (p_2 + q_2)^2 + \dots + (p_n + q_n)^2} \tag{3.1}$$

where $p = (p_1, p_2, \dots, p_n)$ and $q = (q_1, q_2, \dots, q_n)$ are two data points in some dataset X.

Cosine similarity has similar applications as Euclidean distance and can be used to process datasets with a significant number of zeros. The metric value is calculated as the cosine value of the angle between vectors that represent two objects:

$$S_C(A, B) = \frac{A \cdot B}{||A|| ||B||} = \frac{\sum_{i=1}^{n} A_i B_i}{\sqrt{\sum_{i=1}^{n} A_i^2 \cdot \sum_{i=1}^{n} B_i^2}} \tag{3.2}$$

where A and B are n-dimensional vectors of attributes, and A_i and B_i are the ith components of these vectors

The *Jaccard index* is a similarity measure used in clustering techniques, particularly when dealing with binary data, such as presence/absence or Boolean data [22]. However, it can also find wider applications, including in e-commerce customer analysis [23].

It measures the similarity between two finite sample sets as the ratio of the intersection to the union of two sets (A and B):

$$J(A, B) = \frac{A \bigcap B}{A \bigcup B} \tag{3.3}$$

In generalized form, the Jaccard similarity coefficient between two vectors $A = (a_1, a_2, a_3, \dots, a_n)$ and $B = (b_1, b_2, b_3, \dots, B_n)$ is calculated as

$$J(A, B) = \frac{\sum_i \min(a_i, b_i)}{\sum_i max(a_i, b_i)} \tag{3.4}$$

Manhattan distance is a commonly used similarity measure in clustering techniques that work with continuous data, defined as

$$d_T(p, q) = ||p - q||_T = \sum_{i=1}^{n} |p_i - q_i| \tag{3.5}$$

where $p = (p_1, p_2, \dots, p_n)$ and $q = (q_1, q_2, \dots, q_n)$ are two vectors in an n-dimensional real vector space with the fixed Cartesian coordinate system.

Clustering Algorithms

Categorizing clustering algorithms can be challenging due to the potential for overlap between categories, where a method may exhibit features from several categories. A commonly used method for classification involves [44]:

- *Partitioning methods*
- *Hierarchical methods*
- *Density-based methods*
- *Grid-based methods*
- *Model-based methods*
- *Spectral methods*
- *Model evaluation-based methods*

It is worth highlighting that *Model evaluation-based methods* are a set of algorithms that allow the analysis of the quality of other clustering approaches.

Partitioning Methods

The initial assumption for partitioning techniques is the number of clusters. For a dataset consisting of n objects, an algorithm for partitioning divides them into k partitions (where $k \leq n$) where each partition represents a cluster. The objective partitioning criterion—proximity function based on distance—is then optimized to generate clusters where objects are *similar* within each cluster and *dissimilar* to other objects in different clusters.

Partitioning methods include algorithms:

- *K-means*
- *K-medians*
- *Partitioning Around Medoids* (PAM)
- *Clustering Large Applications* (CLARA)
- *Clustering Large Applications based on RANdomized Search* (CLARANS)
- *Fuzzy C-means*

Hierarchical Methods

Hierarchical clustering is a method that organizes data objects into a cluster hierarchy. There are two primary types of hierarchical clustering methods: *Agglomerative* and *Divisive*. They differ in their approaches to combining clusters with objects. Agglomerative methods start with individual data points as separate clusters and progressively merge the most similar clusters until a single macrocluster is formed. In contrast, the divisive approach begins by grouping all data points together into a single macrocluster and then iteratively partitions them into smaller clusters.

Hierarchical methods include algorithms:

- *Single Linkage* (nearest neighbor, K-nearest neighbor—KNN)
- *Complete Linkage* (furthest neighbor)
- *Average Linkage*
- *Ward's Method*
- *Divisive Clustering*
- *Balanced Iterative Reducing and Clustering using Hierarchies* (BIRCH)

- *Clustering Using Representatives* (CURE)
- *RObust Clustering using linKs* (ROCK)

Density-Based Methods
Partitioning and hierarchical techniques typically tend to identify clusters with spherical shapes. In contrast, density-based clustering methods are better suited for detecting clusters with variable shapes (nonspherical) and sizes within a dataset. This detection depends on the local density of data points. These methods are particularly advantageous when dealing with noisy and outlying data. They are well suited for datasets where clusters can exhibit different densities or when the exact cluster count is unknown in advance.

Density-based methods include algorithms:

- *Density-Based Spatial Clustering of Applications with Noise* (DBSCAN)
- *Ordering Points to Identify the Clustering Structure* (OPTICS)
- *Hierarchical Density-Based Spatial Clustering of Applications with Noise* (HDBSCAN)
- *Voronoi-Based Density-Based Spatial Clustering of Applications with Noise* (VDBSCAN)
- *DENsity-based CLUstEring* (DENCLUE)

Grid-Based Methods
Grid-based clustering methods find practical applications in extensive datasets because of their efficiency in dividing the data space into cells and their ability to handle data in a scalable manner. However, one limitation of these methods is their sensitivity to datasets with asymmetrical cluster shapes or varying cluster densities, as the size and shape of grid cells can impact clustering quality. Grid-based clustering methods are commonly used for large datasets, especially those involving spatial data or data with a clear grid structure, such as geographic data or image analysis.

Grid-based methods include algorithms:

- *Statistical Information Grid* (STING)
- *CLustering In QUEst* (CLIQUE)
- *WaveCluster*
- *GRIDSCAN*
- *Density Grid-based Clustering* (DenGRID)
- *STreaming Algorithm for Grid basEd Clustering* (STAGE)

Model-Based Methods
Model-based clustering methods utilize statistical models to partition data into distinct clusters based on a probabilistic model. The fundamental idea is that the data is generated from a combination of various probability distributions, where each distribution represents an individual cluster. The goal is to identify the characteristics of these distributions, such as their averages, variations, and mixing ratios, that best fit the observed data. Model-based clustering methods are particularly valuable in situations where the data distribution is complex and cannot

be separated effectively using simple distance-based techniques. Additionally, they can assist in handling uncertainty analysis and dealing with missing data.

Model-based methods include algorithms:

- *Gaussian Mixture Model* (GMM)
- *Finite Mixture Model* (FMM)
- *Bayesian Gaussian Mixture Model*
- *Hidden Markov Model* (HMM)
- *Dirichlet Process Gaussian Mixture Model* (DPGMM)
- *Expectation-Maximization*(EM)
- *STreaming Algorithm for Grid basEd Clustering* (STAGE)

Spectral Methods
Spectral clustering techniques make use of the eigenvalues and eigenvectors of a similarity matrix or graph representation of the data. This method is particularly advantageous when dealing with intricate datasets that lack well-defined cluster structures or have nonconvex shapes. However, spectral clustering may face scalability challenges when dealing with large datasets because of the computation of eigenvectors and eigenvalues.

Spectral clustering methods include algorithms:

- *Normalized Cut*
- *Ratio Cut*
- *Spectral Clustering by Ng, Jordan, and Weiss*
- *Spectral Clustering by Shi and Malik*
- *Multiclass Spectral Clustering*

Model Evaluation-Based Methods
Model evaluation methods are utilized in cluster analysis to estimate the quality and validity of clustering results generated by various clustering algorithms. The primary goal of these methods is to evaluate the performance of clustering algorithms and assist in the selection of the most suitable clustering solution.

Model evaluation-based methods include algorithms:

- *Silhouette Analysis* (Silhouette Score)
- *Dunn Index*
- *Davies-Bouldin Index*
- *Calinski-Harabasz Index* (Variance Ratio Criterion)
- *Adjusted Rand Index* (ARI)
- *Normalized Mutual Information* (NMI)
- *Purity*
- *Rand Index* (RI)

Selecting the right clustering algorithm can be a challenging task, given the multitude of available methods. It is often difficult to predict which technique will yield the best results, as it depends on the clustering's purpose and the characteristics of the dataset being analyzed. Therefore, when developing a platform to support specialized user interfaces, it is prudent to explore various clustering methods and choose the most suitable one after assessing the quality of clustering and how well it aligns with the business requirements and objectives in the clustering context. Additionally, it is valuable to consider, as one of the selection criteria, the prevalence of specific clustering algorithms in e-commerce solutions.

3.3 Selected Clustering Algorithms Used in E-commerce

A review of the literature concerning the application of clustering in e-commerce demonstrates partitioning methods, notably K-means [24–28], K-mode [29], K-medoids [30–32] and their extension Partitioning Around Medoids (PAM) [33], and Fuzzy C-Means [34]. Commonly used are also other clustering algorithms such as DBSCAN[35], Ward's Method [36, 37], and ROCK [38]. Many aggregate studies compare the features of different clustering approaches and consider the desirability of their application in specific business situations, e.g.:

- In e-commerce product recommendation, comparison of GMM, K-means, DB-SCAN, KNN, etc. [39]
- In cross-border e-commerce, comparison of K-means, DBSCAN [40]
- In movie recommendation, comparison of K-means, BIRCH, DBSCAN [41]
- In e-commerce search engines, comparison of K-means, DBSCAN [42]
- In e-commerce customer analysis, comparison of K-means, DBSCAN, BIRCH [43]

Three clustering algorithms were selected to exemplify different approaches: K-means, BIRCH, and DBSCAN, each representing a distinct group of methods. These choices were made because of their prevalent utilization within the e-commerce industry. When establishing a platform to offer a tailored interface, these algorithms can form the core set of clustering techniques to be implemented.

K-means (Partitioning Method)
K-means is a well-established and commonly used partitioning method that segments data into K distinct clusters to minimize the sum of squared Euclidean distances between data points and their respective cluster centroids within each cluster. The centroid, in this context, serves as the central point of a cluster and is typically defined as the mean (average) of the coordinates of the data points in that cluster.

Formally, it can be said that if a dataset D contains n objects in Euclidean space, the method distributes these objects among k clusters $(C_1, ..., C_k)$ such that $C_i \subset D$ and $C_i \cap C_j = \emptyset$ for $(1 \leq i, j \leq k)$ [44]. The objective function aims for high intracluster similarity and low intercluster similarity. The difference between an object $p \in C_i$ and the centroid c_i, which represents the cluster C_i, is measured by the function $\text{dist}(p, c_i)$, where $\text{dist}(x, y)$ is the Euclidean distance between two points x and y. The quality of cluster C_i is measured by the within-cluster variation, which is the sum of squared error between all objects in C_i and the centroid c_i:

$$E = \sum_{i=1}^{k} \sum_{p \in C_i} dist(p, c_i)^2 \qquad (3.6)$$

where E is the sum of the squared error for all objects in the dataset D, p is the point in space representing an object, and c_i is the centroid of cluster C_i

The objective function tries to generate the resulting k clusters as compact and as separate as possible. The pseudocode of K-means clustering is presented in Appendix A.1.

The clusterization process includes the following steps [44]:

1. *Set the number of clusters*: decision on the number of clusters.
2. *Initialize centroids*: random selection of k of the objects in the dataset as initial centroids.
3. *Assign data points to clusters*: calculation of the distance between each object and each of the k centroids and assignation of this object to the cluster whose centroid was the nearest (using distance metric, e.g., Euclidean distance).
4. *Update centroids*: recalculation of the centroids of the k clusters based on the mean of the objects currently assigned to each cluster.
5. *Repeat steps 3 and 4 until convergence*: iterative repetition of steps until the cluster assignments do not change significantly or until a certain number of iterations is reached.
6. *Verify output*: analysis of the output which is the set of the k clusters, with each object belonging to one of the clusters based on its nearest centroid.

Advantages of K-means clustering:

- *Simplicity*: easy to understand (intuitive) and implement, making it widely used
- *Scalability*: computationally efficient and can handle large datasets efficiently

Limitations of K-means clustering:

- *Sensitiveness to initial centroid positions*: The final result can be sensitive to the initial centroid positions, leading to different solutions for different initializations.
- *Assumption of spherical clusters* may work poorly when clusters have irregular shapes or varying sizes.
- *Predefined number of clusters*: The number of clusters k needs to be defined in advance, which may need several iterations to determine the optimal number of clusters.

Several modifications to the K-means method have been proposed to improve its effectiveness and address its limitations. Some notable variations include K-medoids, PAM, and K-means++. It is important to recognize that while these methods share common assumptions, the clustering results they yield can differ significantly. Additionally, in practice, multiple runs of K-means can be executed with different initialization parameters, and the clustering outcome with the lowest value of the chosen cost function is typically considered the final solution.

BIRCH (Hierarchical Method)

Balanced Iterative Reducing and Clustering Using Hierarchies is a hierarchical clustering algorithm proposed by Tian Zhang, Raghu Ramakrishnan, and Miron Livny in 1996 [1]. BIRCH was designed specifically for clustering large volumes of numeric data. The algorithm combines hierarchical clustering, applied in the initial stage, with other clustering techniques like iterative partitioning, which is used at a later stage [44].

BIRCH uses the concept of *clustering features* to summarize a cluster and the concept of *clustering feature tree (CF-tree)* to represent a cluster hierarchy. The CF tree consists of nodes, where each node represents a cluster prototype and stores information about the clustering features.

Within a dataset that consists of clusters of n d-dimensional data objects (points), the clustering feature (ClF) of the cluster is a 3D vector summarizing information about clusters of objects and can be defined as

$$ClF = (n, LS, SS) \tag{3.7}$$

where LS is the linear sum of the n points $(\sum_{i=1}^{n} x_i)$ and SS is the square sum of the data points $(\sum_{i=1}^{n} x_i^2)$.

A clustering feature is crucial for calculating statistics of a cluster, such as the claster's centroid (x_0), radius—s the average distance from member objects to the centroid (R) and diameter—the average pairwise distance within a cluster (D):

$$x_0 = \frac{\sum_{i=1}^{n} x_i}{n} = \frac{LS}{n} \tag{3.8}$$

$$R = \sqrt{\frac{\sum_{i=1}^{n} (x_i - x_0)^2}{n}} = \sqrt{\frac{nSS - 2LS^2 + nLS}{n^2}} \tag{3.9}$$

$$D = \sqrt{\frac{\sum_{i=1}^{n} \sum_{j=1}^{n} (x_i - x_j)^2}{n(n-1)}} = \sqrt{\frac{2nSS - 2LS^2}{n(n-1)}} \tag{3.10}$$

Such approach allows us to avoid storing detailed information about individual objects or points and use the *clustering feature* only. The pseudocode of BIRCH clustering is presented in Appendix A.2. The main steps of the BIRCH clustering process are:

1. *CF Tree Construction*: Initially, data points are taken from the dataset and inserted into the CF tree to create initial clusters.

2. *CF Tree Maintenance*: When new data points are inserted into the CF tree, its structure is updated using a set of rules to merge or split nodes to maintain a balanced and compact representation.
3. *Clustering Feature Accumulation*: The growing number of data points allows analyzing the data distribution and adaptively adjusts the clusters' boundaries.
4. *Global Clustering*: Once the CF tree has been generated, the algorithm starts global clustering to update the initial clusters using another clustering algorithm (e.g., K-means) on the centroids of the nodes to construct the final clusters.
5. *Hierarchical Representation*: Results can be presented in a hierarchical way allowing users to explore the structure of clusters.

Advantages of BIRCH clustering:

- *Scalability*: It can handle large datasets using a memory-efficient data clustering feature tree to represent the data distribution, which reduces the memory requirements and computational complexity.
- *Efficiency*: Clustering is performed in a single pass through the data, which makes it faster than many other clustering algorithms, especially for large datasets.
- *Incremental learning*: New data points can be easily added to an existing model without recomputing the entire clustering structure.
- *Hierarchical representation* allows users to explore clusters at different levels of granularity and can be useful for understanding the data's structure.

Disadvantages of BIRCH clustering:

- *Sensitiveness to parameter setting*: The set of parameters, such as the maximum number of CFs in a node and the threshold for merging clusters, must be predefined, and the performance of the algorithm can be sensitive to these parameter choices.
- *Assumptions on data distribution*: If the data does not fit well into this assumption, the clustering performance may be low.
- *Limited to spherical clusters* works well for compact and spherical-shaped clusters, but it may struggle with irregularly shaped or elongated clusters.
- *Handling of outliers* which can negatively impact the clustering results, especially in datasets with noisy or sparse data
- *Lack of cluster interpretation* may not provide meaningful cluster labels or interpretations which may require additional analysis to understand the characteristics of each cluster.

DBSCAN (Density-Based Method)

In 1996, Martin Ester, Hans-Peter Kriegel, Jörg Sander, and Xiaowei Xu proposed the *density-based spatial clustering of applications with noise* algorithm [46]. This algorithm does not make any assumptions regarding the number and shape of the clusters. DBSCAN is a nonparametric algorithm that employs density-based clustering to group closely located points (those with many neighboring points), while flagging as outliers the points in low-density regions (whose nearest neighbors are too far away). The pseudocode of the DBSCAN clustering is presented in Appendix A.3.

The DBSCAN algorithm has two main hyperparameters:

- *epsilon (ϵ)* defines the maximum distance between two points for them to be considered neighbors and determines the radius of the neighborhood around each point.
- *min_points* specifies the minimum number of points required to form a dense region or cluster.

The main steps of the DBSCAN clustering process are [47]:

1. *Core points*: A data point is considered a core point if there are at least *min_points* data points within its ϵ-neighborhood, core points are the central building blocks of clusters.
2. *Expand clusters*: Starting from a randomly chosen core point, the algorithm expands the cluster by adding its neighbors to the cluster, and if a neighbor is also a core point, its neighbors are recursively added to the cluster.
3. *Border Points*: These are points that are within distance ϵ from a core point but do not have enough neighbors to be a core point are labeled as *border points* and are included in the cluster of their corresponding core point but do not participate in expanding the cluster further.
4. *Noise Points*: These are data points that are not core points or border points are considered *noise points* or *outliers* and are not assigned to any cluster.

The resulting clusters may vary in shape and size, and DBSCAN can successfully detect clusters despite the existence of noise and outliers.

Advantages of DBSCAN clustering:

- *Ability to find clusters of arbitrary shape*: It can identify clusters of various shapes and sizes, making it suitable for datasets where traditional methods might struggle due to assumptions about cluster shapes.
- *Robust to noise and outliers*: It can effectively handle noisy data points and outliers, as they are considered as noise and not assigned to any cluster.
- *Insensitive to the order of data*: No dependency on the order of the data points, which makes the method less sensitive to the initial configuration of the data.
- *Minimum input parameters*: There are only two parameters that can be chosen based on domain knowledge or using heuristic methods
- *No predefined number of clusters*: It does not require specifying the number of clusters beforehand and automatically determines the number of clusters based on the data and the chosen parameters.
- *Efficiency*: Time complexity is relatively low, making this method suitable for clustering large datasets.

Limitations of DBSCAN clustering:

- *Sensitiveness to parameter tuning*: Poor parameter choices can lead to suboptimal or incorrect clustering results—if ϵ is too small, it may lead to many data points being considered outliers, while if it is too large, it may merge different clusters into one.

- *Not suitable for high-dimensional data*: In high-dimensional spaces, the notion of density becomes less meaningful, which can negatively impact the performance.
- *Difficulty with varying density clusters*: In cases where the density of clusters varies significantly, some areas with lower densities might be classified as outliers or noise.
- *Border point ambiguity*: Points on the border between two clusters might be assigned to either cluster, leading to some ambiguity in the clustering results for certain data points.
- *Computational complexity*: Complexity can increase significantly for spatial databases due to the need for range queries when calculating neighborhood points.

The selection of a clustering algorithm plays a pivotal role in e-commerce customer segmentation [48]. The choice of algorithm determines which groups of users will receive a designated interface variant. Moreover, the method employed shapes the characteristics of each cluster and its users, thereby influencing behavioral variances within the identified customer groups. An analysis of the specific clusters generated and the characteristic behavior of the customers assigned to them is an integral part of preparing the information necessary to support dedicated UI variants.

At this point, it is worth noting that among the papers included in the analysis of academic publications on the use of clustering algorithms for customer segmentation in e-commerce published between 2000 and 2022 [34], there is not a single one that refers to applications of this approach to the design of personalized user interfaces. This shows that the issue addressed has not yet been studied in depth and, in this context, represents a research gap worth filling.

3.4   Customer Cluster Analysis

When preparing a dedicated interface for e-shop customers clustered in groups, a human expert needs to identify the key differences between clusters. These differences serve as the basis for selecting modifications that create interface variants specific to each group.

The analysis of differences between clusters can cover two aspects:

- Values of the parameters characterizing the clusters
- Analysis of customer journey in an e-shop

Parameters Describing the Clusters
The effect of clustering, apart from dividing a set of customers into groups, can be to calculate the values of various types of indicators relating to the designated cluster. These may relate to the quality of clustering (e.g., cluster size or cohesion) and to the behavior of users assigned to the cluster.

	0	1	2	3
size	14870	18080	9966	5873
size proc	30.4782%	37.0575%	20.4267%	12.0375%

Fig. 3.3 Example of clustering result with cluster size

The first group of indicators focuses on assessing the number of clusters, a critical factor in developing interface alternatives. A lack of clusters may render a dedicated interface ineffective, while an overwhelmingly dominant cluster might make a multivariant interface unnecessary. Evaluating the cluster count is particularly important for methods like DBSCAN, which do not predefine the number of resulting clusters. This relevance arises from the need to design dedicated user interfaces for these clusters, which can become challenging with a large number of clusters (Fig. 3.3).

The minimum cluster size (e.g., X% of the total customer population) or the variation in cluster size (e.g., the variance in the size of the resulting clusters) can serve as the acceptance criterion in this instance. Furthermore, analyzing cluster counts can provide insight into the preferred order of design of dedicated interface variants—starting with the most numerous clusters and progressing to the least numerous.

Verifying the internal consistency and similarity of clusters is a more intricate task. Several methods can assess the quality of clustering algorithms based on internal criteria, including:

- *Silhouette Analysis*, which measures an object's similarity to its cluster compared to other clusters
- *Dunn Index*, used to identify sets of clusters that are compact, with small cluster member variance and well separated
- *Davies–Bouldin Index*, which defines the ratio of within-cluster scatter to between-cluster separation

Based on cluster quality indicators, adjustments to clustering parameters, particularly the number of clusters (k) in predefined methods, can be made to optimize the results. However, it is important to note that increasing k may yield better clustering indicators, but it does not necessarily align with improved business context indicators. An increase in the number of clusters also means a reduction in the number of objects assigned to each cluster and a greater need to design dedicated UI variants. Therefore, conducting a preliminary comparative analysis of various clustering methods, considering different essential to determine the optimal configuration that balances high-quality clustering with appropriate cluster sizes, utilizing methodologies such as T-distributed Stochastic Neighbor Embedding (TSNE) [49] or Uniform Manifold Approximation and Projection (UMAP) [50] (Fig. 3.4).

The second set of factors used to describe clusters involves measuring certain characteristics of the objects assigned to them. For example, with e-shop customers,

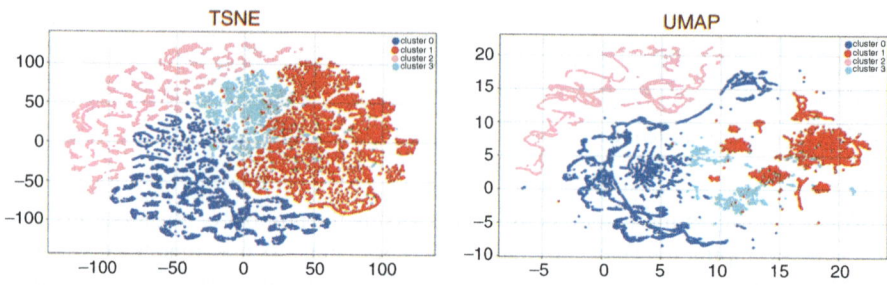

Fig. 3.4 Example of cluster visualization

	label 0, mean/mode	label 1, mean/mode	label 2, mean/mode	label 3, mean/mode
action	17.5074	93.7845	15.3367	36.7037
event	3.1198	22.7626	4.3662	9.5326
firstTimestamp	2023-05-18 10:22:06	2023-05-15 12:29:39	2023-05-16 18:52:36	2023-05-15 16:39:13
lastTimestamp	2023-05-21 04:37:33	2023-05-22 07:25:49	2023-05-17 14:07:25	2023-05-20 10:34:51
revenue	0.0844	10.2206	0.4798	2.5689

Fig. 3.5 Example of clustering result with descriptive factors (mean/mode)

	label 0, std	label 1, std	label 2, std	label 3, std
action	39.3773	139.93	14.3868	49.2022
event	12.4599	38.2872	4.5591	11.3101
firstTimestamp	1310390.6652	1305913.2822	1344243.3278	1311903.7448
lastTimestamp	1302927.1381	1297804.9545	1349102.3638	1308355.2846
revenue	1.8298	34.282	6.4101	11.2657

Fig. 3.6 Example of clustering result with descriptive factors (std. deviation)

this may include the number of actions taken, events attended, and the value of purchases made, among other things (Fig. 3.5).

By examining metrics describing customers assigned to different clusters, it is possible to assess their level of activity, frequency of use of the search engine, as well as the colors, sizes, and numbers and value of the orders they place. Supplementary data may also comprise the variation of specific features (Fig. 3.6).

Examples of conclusions that can be drawn from the analysis of the values of indicators describing customers in clusters can be:

- *If users rarely use the search engine, encourage them to do so by making the search area more visible.*
- *If users have a large number of actions but rarely buy, the purchase path should be shortened.*
- *If users use certain filters, they should be expanded and shown first.*
- *If users have little action, they should be encouraged to stay on the site.*
- *If users have a lot of abandoned shopping carts, consider making checkout easier.*

The conclusions drawn in this way can form the basis for a set of potential user interface changes. The preparation of different options is important at this stage, while their verification and business relevance come later.

Analysis of Customer Journey

The recorded data of customer activity in an Internet shop can be reconstructed into a *path* of their visit. By identifying the patterns of such *journeys*, it is possible to point out areas that require interface adjustments. The journey can be presented as a directed graph, where the points represent recorded e-commerce events and the edges represent the activities completed by the customer.

Analysis of customer journey within the cluster may include:

- Verification of the most frequent events
- Verification of the most frequent activities (transitions between events)
- Verification of the most common activity sequences (series of consecutive activities)

The results of the analysis can be cluster-specific events (e.g., *cluster customers often use a search engine*), activities (e.g., *cluster customers often move from a product card to a minicart*), or sequences (e.g., *cluster customers often move between cards of different products and then put the product in the basket*).

Based on the analysis of customers from successive clusters, it is possible to learn about their specific behavior within the cluster and also to compare it with the behavior of customers from other clusters. Thanks to this—and to the knowledge of UX experts—it is possible to identify changes that need to be made to the interface variants offered to the customer groups studied.

Collecting data on e-shop customer behavior, processing it to identify groups with similar characteristics, and analyzing the differences between the groups are the three steps in preparing the information needed to develop dedicated UI variants. These activities allow a better understanding of customer behavior and tailor user interface changes to their preferences, choices, and decisions. However, this knowledge alone is not enough—the technical capability to design, implement, deploy, and monitor different user interface variants must also be in place.

References

1. The Open Group (1997) DCE 1.1: Authentication and Security Services. https://pubs.opengroup.org/onlinepubs/9696989899/chap1.htm. Cited 15 Jul 2023
2. International Organization for Standardization (2014) Information technology—Procedures for the operation of object identifier registration authorities—Part 8: Generation of universally unique identifiers (UUIDs) and their use in object identifiers. https://www.iso.org/standard/62795.html. Cited 15 Jul 2023
3. Mathis FH (1991) A generalized birthday problem. SIAM Rev. https://doi.org/10.1137/103305
4. G2 (2023) Best Tag Management Systems. https://www.g2.com/categories/tag-management-systems. Cited 15 Jul 2023
5. Alhlou F, Asif S, Fettman E (2016) Google tag manager concepts. In: Google analytics breakthrough: from zero to business impact. https://doi.org/10.1002/9781119266365.ch5
6. Weber J (2015) Tracking interactions with Google Tag Manager. In: Practical Google analytics and Google Tag Manager for developers. https://doi.org/10.1007/978-1-4842-0265-4_5

7. TechTarget Contributor (2019) Definition: tag management system. https://www.techtarget. com/whatis/definition/tag-management-systems-TMS. Cited 15 Jul 2023
8. W3TECHS (2023) Usage statistics of traffic analysis tools for websites. https://w3techs.com/ technologies/overview/traffic_analysis. Cited 24 Jul 2023
9. Alby T (2023) Popular, but hardly used: has Google Analytics been to the detriment of Web Analytics? In: WebSci '23: 15th ACM web science conference 2023. https://doi.org/10.1145/ 3578503.3583601
10. Lubowicka K (2023) Is Google Analytics (3 & 4) GDPR-compliant? [Updated]. https://piwik. pro/blog/is-google-analytics-gdpr-compliant/. Cited 24 Jul 2023
11. Cookiebot Blog (2022) Google Analytics and the GDPR. https://www.cookiebot.com/en/ google-analytics-gdpr/. Cited 24 Jul 2023
12. Quintel D, Wilson R (2020) Analytics and privacy. Inf Technol Libr. https://doi.org/10.6017/ ital.v39i3.12219
13. Wang G, Zhang X, Tang S, Wilson C, Zheng H, Zhao BY (2017) Clickstream user behavior models. ACM Trans Web. https://doi.org/10.1145/3068332
14. Wen Z, Lin W, Liu H (2023) Machine-learning-based approach for anonymous online customer purchase intentions using clickstream data. Systems. https://doi.org/10.3390/systems11050255
15. Kukar-Kinney M, Scheinbaum AC, Orimoloye LO (2022) A model of online shopping cart abandonment: evidence from e-tail clickstream data. J Acad Mark Sci. https://doi.org/10.1007/ s11747-022-00857-8
16. Zavali M, Lacka E, de Smedt J (2023) Shopping hard or hardly shopping: revealing consumer segments using clickstream data. IEEE Trans Eng Manag. https://doi.org/10.1109/TEM.2021. 3070069
17. Ulitzsch E, Ulitzsch V, He Q (2022)A machine learning-based procedure for leveraging clickstream data to investigate early predictability of failure on interactive tasks. Behav Res. https://doi.org/10.3758/s13428-022-01844-1
18. Everitt BS, Landau S, Leese M, Stahl D (2011) Cluster analysis. John Wiley & Sons, London
19. Halkidi M, Batistakis Y, Vazirgiannis M (2001) On clustering validation techniques. J Intell Inf Syst. https://doi.org/10.1023/A:1012801612483
20. Kondruk NE, Malyar MM (2021) Analysis of cluster structures by different similarity measures. Cybern Syst 57. https://doi.org/10.1007/s10559-021-00368-4
21. Mehta V, Bawa S, Singh J (2020) Analytical review of clustering techniques and proximity measures. Artif Intell Rev. https://doi.org/10.1007/s10462-020-09840-7
22. Chung NC, Misasojedow B, Startek M, Gambin A (2019) Jaccard/Tanimoto similarity test and estimation methods for biological presence-absence data. BMC Bioinform. https://doi.org/10. 1186/s12859-019-3118-5
23. Verma V, Aggarwal RK (2020) A comparative analysis of similarity measures akin to the Jaccard index in collaborative recommendations: empirical and theoretical perspective. Soc Netw Anal Min. https://doi.org/10.1007/s13278-020-00660-9
24. Sari P, Purwadinata A (2019) Analysis characteristics of car sales in e-commerce data using clustering model. J Data Sci Appl. https://doi.org/doi.org/10.21108/jdsa.2019.2.19
25. Du X, Liu B, Zhang J (2019) Application of business intelligence based on big data in e-commerce data analysis. J Phys. Conf Ser. https://doi.org/10.1088/1742-6596/1395/1/012011
26. Aria R R (2020) K-means to determine the e-commerce sales model in Indonesia. IJISTECH. https://doi.org/10.30645/ijistech.v3i2.47
27. Cui H, Niu S, Li K, Shi C, Shui S, Gao Z (2021) A K-means++ based user classification method for social e-commerce. Intell Autom Soft Comput. https://doi.org/10.32604/iasc.2021. 016408
28. Li Y, Qi J, Chu X, Mu W (2023) Customer segmentation using K-means clustering and the hybrid particle swarm optimization algorithm. Comput J. https://doi.org/10.1093/comjnl/ bxab206
29. Kamthania D, Pawa A, Madhavan S (2018) Market segmentation analysis and visualization using K-mode clustering algorithm for e-commerce business. J Comput Inf Technol. https:// doi.org/10.20532/cit.2018.1003863

30. Rahardja U, Hariguna T, Baihaqi WM (2019) Opinion mining on e-commerce data using sentiment analysis and K-medoid clustering. In: 2019 twelfth international conference on Ubimedia computing (Ubi-Media). https://doi.org/10.1109/Ubi-Media.2019.00040

31. Peng Q, Zhang S, Zhang J, Huang Y, Yao B, Tang H (2021) Ball K-medoids: faster and exacter. In: Advances in artificial intelligence and security. https://doi.org/10.1007/978-3-030-78615-1_16

32. Wu Z, Lingmin J, Jiali Z, Lizheng J, Liang C (2022) Research on segmenting e-commerce customer through an improved K-medoids clustering algorithm. Comput Intell Neurosci. https://doi.org/10.1155/2022/9930613

33. Gaikwad D, Lamkuche H (2021) Segmentation of services provided by e-commerce platforms using PAM clustering. J Phys Conf Ser. https://doi.org/10.1088/1742-6596/1964/4/042036

34. Wang L, Jiang Y (2022) Collocating recommendation method for e-commerce based on fuzzy C-means clustering algorithm. J Math. https://doi.org/10.1155/2022/7414419

35. Yang Y, Jiang J, Wang H (2014) Application of e-commerce sites evaluation based on factor analysis and improved DBSCAN algorithm. In: 2014 international conference on management of e-commerce and e-government. https://doi.org/10.1109/ICMeCG.2014.17

36. Triandini E, Hermawati FA, Ketut I (2020) Hierarchical clustering for functionalities e-commerce adoption. Jurnal Ilmiah KURSOR. https://doi.org/10.21107/kursor.v10i3.230

37. Scutariu A-L, Şuşu Ş, Huidumac-Petrescu C-E, Gogonea R-M (2022) A cluster analysis concerning the behavior of enterprises with e-commerce activity in the context of the COVID-19 pandemic. J Theor Appl Electron Commer Res. https://doi.org/10.3390/jtaer17010003

38. Tian H (2023) Clustering and analysis of rural e-commerce live broadcast mode based on data orientation. Int J Comput Intell Syst. https://doi.org/0.1007/s44196-023-00269-8

39. Wang K, Zhang T, Xue T, Lu Y, Na S-G (2019) E-commerce personalized recommendation analysis by deeply-learned clustering. J Vis Commun Image Represent. https://doi.org/10.1016/j.jvcir.2019.102735

40. Zhao Y, He Y, Wutao Z (2022) Personalized clustering method of cross-border e-commerce topics based on ART algorithm. Math Probl Eng. https://doi.org/10.1155/2022/8190544

41. Nawara D, Kashef R (2021) Deploying different clustering techniques on a collaborative-based movie recommender. In: 2021 IEEE international systems conference (SysCon). https://doi.org/10.1109/SysCon48628.2021.9447139

42. Darwin, Purba R, Pasha MF (2020) Search query clustering comparation on e-commerce using K-means and adaptive DBSCAN. In: 2020 3rd international conference on mechanical, electronics, computer, and industrial technology (MECnIT). https://doi.org/10.1109/MECnIT48290.2020.9166610

43. Jabade V, Ghadge S, Jamadar M, Girase P (2023) Customer segmentation for smooth shopping experience. In: 2023 4th international conference for emerging technology (INCET). https://doi.org/10.1109/INCET57972.2023.10170126

44. Han J, Kamber M, Pei J (2011) Data mining: concepts and techniques. Elsevier, Amsterdam

45. Saputra MF, Widiyaningtyas T, Wibawa A (2018) Illiteracy classification using K means-naïve Bayes algorithm. Int J Inform Vis. https://doi.org/10.30630/joiv.2.3.129

46. Ester M, Kriegel H-P, Sander J, Xu X (1996) A density-based algorithm for discovering clusters in large spatial databases with noise. In: Proceedings of the second international conference on knowledge discovery and data mining (KDD-96)

47. Schubert E, Sander J, Ester M, Kriegel H-P, Xu X (2017) DBSCAN revisited, revisited: why and how you should (Still) Use DBSCAN. ACM Trans Database Syst. https://doi.org/10.1145/3068335

48. Wasilewski A (2023) Clusterization methods for multi-variant e-commerce interfaces. Ann Comput Sci Inform Syst. https://doi.org/10.15439/2023F1377

49. Hinton GE, Roweis ST (2002) Stochastic neighbor embedding. NIPS. https://api.semanticscholar.org/CorpusID:20240. Cited 24 Jul 2023

50. McInnes L, Healy J, Melville J (2020) UMAP: uniform manifold approximation and projection for dimension reduction. https://arxiv.org/pdf/1802.03426.pdf. Cited 24 Jul 2023

Chapter 4
Designing and Serving a Dedicated Interface

4.1 Designing a Dedicated Interface Variant

A *user interface variant* is a version or variation of a UI design that has been modified in some way from the original or main design, which can be described as a default *user interface*. UI variants may be created to suit different contexts, devices, user needs, or design goals. They represent a tailored version of the UI designed to meet specific requirements, whether these are related to technical constraints, user preferences, or business goals.

Common scenarios of UI variants using include:

- **Responsive Design**: In web-based solutions, a responsive UI variant is designed to adapt the interface to different screen sizes and orientations, ensuring that the user interface remains usable and visually appealing on devices ranging from desktop monitors to smartphones and tablets.
- **Localization**: Variants created to accommodate different languages, cultures, regions, cultural considerations, and legal requirements that may necessitate adjustments to the layout or content provided to the customer (e.g., omission of content or images prohibited by law)
- **Platform-Specific Design**: Variants created to conform to the design guidelines and conventions of different software development platforms (e.g., iOS, Android, and desktop) or browsers (e.g., Chrome, Opera, and Safari) to ensure a consistent and intuitive user experience for users across devices
- **Accessibility Design**: User interfaces designed to meet accessibility needs, such as variants that accommodate different levels of visual impairment, allowing users to adjust text size, color contrast, or navigation methods
- **Functional Design**: Variants created to support different sets of features or functionalities, e.g., a simplified UI variant for users who only need basic features (or have lower access rights), while another variant includes more advanced options

A. Wasilewski, *Multi-variant User Interfaces in E-commerce*, Progress in IS, https://doi.org/10.1007/978-3-031-67758-8_4

- **Personalization**: Personalized UI variants based on user preferences or behavior, allowing the layout, color scheme, or placement of UI elements to be adapted to user preferences

Another common use case for interface variants is A/B testing [1], where different UI variants are tested simultaneously to determine which one is more effective at achieving specific goals. This approach can also be applied to analyze the effectiveness of personalized interface variants.

User interface variants can generally be categorized as *static* and *dynamic*. *Static variants* are designed and implemented with specific UI versions and served to designated audiences. This approach allows for the swift delivery of a multivariant interface to well-defined user groups. For instance, in e-commerce, this could involve dedicated user interfaces for B2B and B2C customers within a single sales platform. The advantage of this method is the ability to create independent and distinctly different interface variants. However, its drawback is limited flexibility in terms of interface modifications (subsequent changes require design and implementation) and user grouping (a fixed number of groups with predetermined characteristics, corresponding to the number of interface variants, and a relatively simple algorithm for assigning customers to groups).

Dynamic variants represent a more complex solution during the design and implementation stage. However, they offer a high level of flexibility in altering the appearance of interfaces and tailoring them to user groups with characteristics that may not have been known during the UI design stage. This approach becomes essential when a multivariate user interface is to be delivered to e-commerce customer groups based on an analysis of their behavior.

Preparation of dynamic interface variants includes the following activities:

- Designing the default interface
- Division of the views into areas
- Planning potential modifications in areas
- Implementation of modifications

The foundation for creating dynamic e-commerce interface variants is a default (basic) interface, which can be customized to meet the specific requirements and preferences of distinct customer groups. The core interface design should adhere to the rules of UI development for e-commerce solutions, treating all users as a unified group. Although this approach may require compromises, it leverages knowledge of potential options for interface elements to facilitate the design of a multivariant interface. The UX approach plays a pivotal role in this process. A positive user experience can result in higher conversion rates, increased customer retention, and an enhanced brand reputation [2]. Improving UX can be guided by principles derived from the ISO 9241-11 standard to enhance user satisfaction [3].

In the context of e-commerce, there are several fundamental views (web pages) that can undergo interface variations. These primarily include the home page, product listing, product card, shopping cart, and checkout, among other elements of the online store whose appearance can influence customers' purchasing behavior.

Fig. 4.1 Dividing the listing view into areas (https://4fstore.com/)

Each of these views can possess a distinct look and feel. To streamline the management of interface variants, it is beneficial to *divide each view* into sections within which appearance modifications will be designed. An example of segmenting the listing view into sections is illustrated in Fig. 4.1.

Within each area, a number of modifications can be made to create the different interface variants. Examples of modifications for each area are shown in Table 4.1.

Table 4.1 Examples of UI modifications

Area ID	Area name	Modification	ID
AREA 1	Top menu	The order of categories	1-1
		Additional virtual categories	1-2
AREA 2	Search engine	Hidden input field	2-1
		Large magnifier icon	2-2
		Small magnifier icon	2-3
		Additional description in the input field	2-4
AREA 3	User management	Different order of icons	3-1
		Enlarge login icon	3-2
AREA 4	Side menu	Shown	4-1
		Hidden	4-2
		Different level of category depth	4-3
AREA 5	Listing parameters	Swap places of sorting and selecting the number of products	5-1
		Changed default parameter values	5-2
AREA 6	Product box	Possibility to choose product variant	6-1
		Add to cart icon	6-2
		Price highlight	6-3
AREA 7	Filters	Changing the order of filters	7-1
		Collapsed selected filters	7-2
		Expand selected filters	7-3

Individual modifications within an area may also coexist (e.g., modifications 2-1 and 2-2), leading to the creation of additional combinations that are available for implementation (Table 4.2) with their distinct identifiers (e.g., 2–11 to 2–14). Consequently, it is feasible to generate a substantial number of interface variants by choosing various combinations of modifications.

The potential changes listed are not exhaustive; their size and scope depend on the perceived potential of UX specialists. Similar change lists should be prepared for all views with UI variants. These lists may expand over time as new ideas emerge on tailoring the UI to customer expectations. Conversely, some modifications may be removed if they are confirmed to not affect UI effectiveness or negatively impact customer behavior or decisions.

Based on the defined areas and the available modifications within the areas, an interface variant can be configured as a combination of area identifiers and modification identifiers. The definition of an interface variant can be represented in

Table 4.2 Options of the search input field modifications

Appearance				Description	ID
ANA		Q	PIE	Hidden input field for writing, field only visible after clicking on magnifying glass	2-11
ΛNA	Meklēt veikalā...	Q	PII	Input field shown, small magnifying glass	2-12
ANA		Q	PIE	Hidden input field, magnifying glass icon enlarged (25x25px)	2-13
ΛNA	Meklēt veikalā...	Q	PII	Input field shown, magnifying glass icon enlarged (25x25 px)	2-14

JSON format and submitted to the browser as a view configuration to be displayed to a specific customer, e.g.:

```
Interface Variant Definition
{
    "AREA2":"2-12"
    "AREA3":"3-1"
    "AREA5":"5-2"
    "AREA6":"6-1"
    }
```

When assigning an interface variant to clusters of e-shop customers, there is no need to send a detailed variant definition to the browser. Providing the identifier of the predefined interface variant is sufficient, as it will be the same for all customers in the cluster. Interface variant definitions can be stored in the e-shop, and the recommendation part would link the specific UI variant to the customer group resulting from clustering.

4.2   UI Variants and Customer Groups

If one wants to design variants of interfaces for groups of customers resulting from clustering, it is necessary to choose the available modifications in such a way that they best match the behavior of users. As there is a list of areas and modifications on

	label 0, mean/mode	label 1, mean/mode	label 2, mean/mode	label 3, mean/mode	label 4, mean/mode
action	17.3778	59.6057	16.0189	179.0798	41.4345
event	3.1096	13.9615	4.6503	43.8591	11.0128
revenue	0.0861	3.047	0.4766	24.6469	4.637
ecommerceOrder	0.0031	0.062	0.0078	0.3926	0.1079
search	0.1759	0.4495	0.1775	1.0463	0.3207

Fig. 4.2 Excerpt from the clustering report—average values

	label 0, std	label 1, std	label 2, std	label 3, std	label 4, std
action	36.279	79.9747	14.3574	190.1289	48.3883
event	11.3439	26.6866	4.6226	53.1931	12.3219
revenue	1.8716	15.8081	6.3185	51.903	18.2802
ecommerceOrder	0.0554	0.2684	0.0894	0.6363	0.3467
search	1.3008	3.5586	1.0163	4.4855	1.5537

Fig. 4.3 Excerpt from the clustering report—standard deviation

the one hand and customer characteristics in the clusters on the other, it is necessary to link them based on the identified characteristics of each user group.

Knowledge about the behavior of customer groups can be obtained from two sources:

• A set of values of characteristics describing clusters
• Knowledge of actions and sequences of actions taken by customers from a specific cluster

The basic characteristics of users included in specific clusters can be presented as part of the clustering report. Analysis of means (Fig. 4.2) and standard deviations (Fig. 4.3) provides basic information about customer behavior.

The list of characteristics included in the report can include predefined and additional characteristics that are key to customer similarity within clusters and differentiation between clusters.

Analyzing clustering reports allows considering characteristics such as the number of actions and events, indicating the activity of using the online shop, purchases made (i.e., revenue for the shop owner), frequency of search engine use, the most frequently selected product categories and their attributes (e.g., color or size), use of descriptions and photos on the product card, etc. The conclusions drawn on this basis are general but allow tentatively planning an in-depth analysis of customer behavior. This helps decide which customer clusters should be studied in detail and for which it is worth designing a dedicated interface variant. For clusters containing customers with low e-commerce activity or low sales, designing and serving a dedicated user interface may not be economically viable.

Reports on actions and action sequences can form the basis for an in-depth analysis of customers assigned to specific clusters. Actions encompass any type of user activity that changes the session's state in the e-shop, such as jumping to a selected page, selecting a category or product, using specific filters or a search engine, displaying a product feature, and pressing navigation buttons. Analyzing

action	actionsOccurrencesCount	uniqueCustomerTokensCount
homepage	22073	4511
other	7512	1921
listing-izpardosana	6661	1721
account-customer-account	4685	469
listing-izpardosana-sievietes	3790	987
account-customer-account-login	3083	504
listing-kategorijas-back-to	2763	675
listing-izpardosana-viriesi	2289	573
event-Search-Click	1672	960
listing-sievietes-apgerbs-jakas-un	1623	389
listing-sievietes	1516	415
listing-izpardosana-sievietes-apgerbs	1314	368
other-izpardosana-zeniem-izmers-sm_52,56cm	1011	76
listing-viriesi	988	264
listing-izpardosana-zeniem	953	258
listing-viriesi-apgerbs-jakas-un	947	220
landing-landing-hot-deals	941	596
listing-sievietes-apgerbs-peldkostimi	858	223
listing-izpardosana-sievietes-page-2	841	254
listing-izpardosana-meitenes	760	211
listing-izpardosana-viriesi-apgerbs	726	180
account-customer-account-createPassword	669	76
checkout-checkout-cart	651	133
listing-viriesi-apgerbs-krekli-un	622	152
other-sales-order-history	610	116

Fig. 4.4   Example of an action frequency report

frequency (Fig. 4.4) helps identify the most frequently performed actions and actions not performed by customers from a particular cluster. The actions report can be cumulative (the total number of specific actions performed by customers from the analyzed cluster—*actionsOccurencesCount*), or related to unique users from the cluster who performed a specific action—*uniqueCustomerTokensCount*.

Based on such a report, it is possible to address modifications within the interface variant dedicated to customers from the analyzed cluster properly. These modifications should be related to the most frequent actions, while modifications affecting areas that users do not utilize can be ignored.

However, information on the most common actions taken by customers may not be sufficient to analyze their behavior. Sequences, defining a series of consecutive events resulting from the customer journey in the e-shop, are also important. Sequences of different lengths can be analyzed, e.g., a sequence of three elements containing *predecessor—action—successor* (Fig. 4.5). Longer sequences can also be considered, providing a more detailed view of specific customer behavior patterns.

Action sequence reports complement the knowledge about the e-shop usage of customers assigned to specific clusters. Based on these reports, UX specialists can assess which elements of the user interface are relevant and worth modifying. They can also reject changes related to activities that are not performed by customers from the cluster under study.

Another use of sequence reports is to verify the expected conversion path of the e-shop user. If a shop owner wants to present specific information to the customer (e.g., about promotions, loyalty programs, quality of goods on sale, etc.) beyond the standard shopping path, a sequence of actions can be used to assess whether the current UI is effective and meets the objectives set. Checking the consistency of

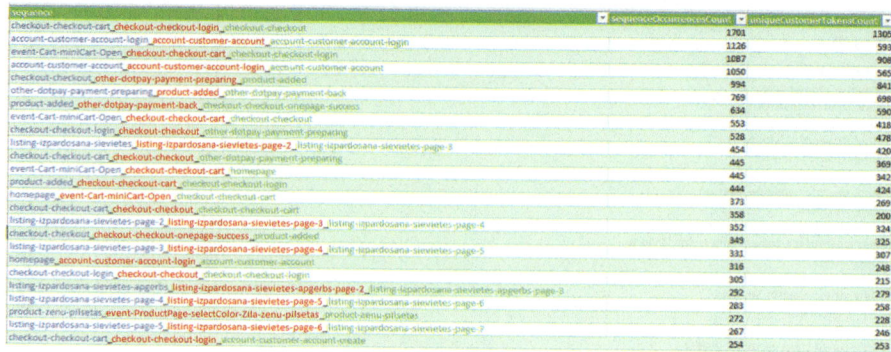

Fig. 4.5 Example of a sequences frequency report

customer behavior with the seller's expectations is one of the metrics that allows the analysis of the performance of the user interfaces served.

4.3   Performance Metrics for User Interfaces

Efficiency metrics in e-commerce play a crucial role in evaluating the performance of an online business. These metrics aid in optimizing operations, enhancing the customer experience, and ultimately increasing profitability. By monitoring these indicators, e-commerce companies can pinpoint areas for future development.

While numerous metrics can be defined to assess various aspects of e-commerce performance, a study by Statista [4] highlights several fundamental metrics used by retailers to measure the effectiveness of personalization initiatives (Fig. 4.6).

Conversion Rate

The most popular metric is the *conversion rate (CR)*, defined as the percentage of users who take a desired action.

The prevailing assumption is that CR represents the percentage of website visitors who purchase on the site. However, in practice, it can encompass any action relevant to the business. Beyond purchasing in an e-shop, the analysis may include activities such as user registration, subscription to content (paid or free), downloading materials (e.g., software, whitepapers) available on the site, utilizing a specific feature of an application, or reading a particular web page, among others. In essence, this metric can be applied to analyze any action that can be unambiguously counted by a computer and aligns with the goals set by the e-commerce owner.

In general, the method of calculating the CR ratio is relatively simple and can be expressed by the following formula:

$$CR = \frac{C_o}{I} \tag{4.1}$$

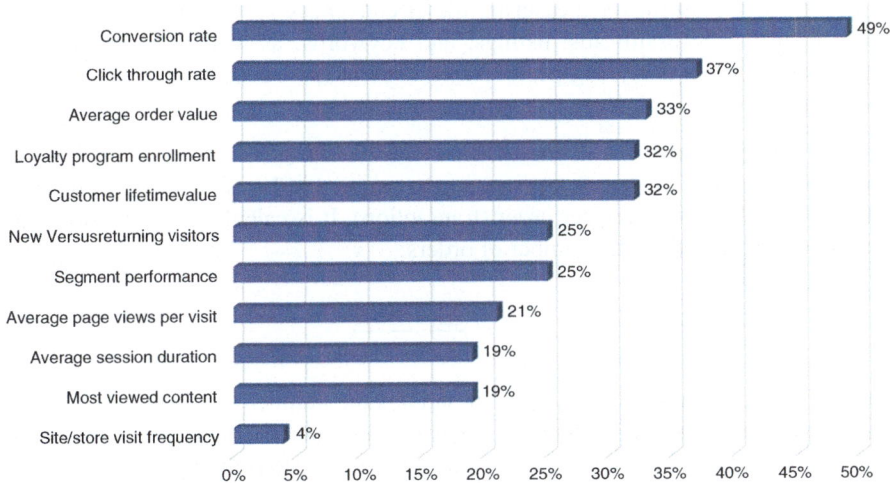

Fig. 4.6 The primary metrics used by retailers to measure personalization initiatives [4]

where C_o is the number of conversions and I is the number of interactions.

In the context of a conversion rate related to sales, C_o can mean the number of orders placed and I the number of user sessions in the online shop over a given period. This approach to the CR indicator is crucial from the perspective of e-commerce efficiency, but it possesses a fundamental flaw—it pertains to the completion of the entire purchase process and does not permit a detailed analysis of the path taken by the customer. It can be considered a kind of *macroconversion* .

When optimizing and adapting the UI to user behavior, *microconversion* should also be taken into account, as the vast majority (even more than 97% [8]) of online store traffic does not result in a purchase (macroconversion). Still, it can be crucial for UI personalization. Basic events, such as clicking on a link, watching a video, or scrolling through photos in a gallery, can be used to calculate microconversions. These actions are valuable for UX-oriented website analytics aiming to track smaller design elements [9].

Click-Through Rate
Another important indicator from the Statista survey is the *click-through rate (CTR)*, which shows how often users who see the ad or other information end up clicking on it. This metric can be used to measure how well e-marketing activities are working.

It can be written as

$$CTR = \frac{Cl}{S} \qquad (4.2)$$

where Cl is the number of clicks that the analyzed ad receives and S is the number of times this ad is shown.

This metric is useful for measuring the effectiveness of e-marketing activities. It can identify successful ads, listings, and keywords, as well as areas needing improvement, when analyzing personalized content (products, banner ads, etc.) in an e-shop.

Average Order Value

The third important metric is *average order value (AOV)*, denoting the average amount spent by customers per order (transaction). It is calculated by dividing the total order value by the total number of orders:

$$AOV = \frac{\sum_{i=1}^{n} OV_i}{n} \tag{4.3}$$

where OV_i is the amount of the ith order and n is the number of orders.

The calculation of the AOV index involves some inconveniences related to the life cycle of an order in an e-shop. The main problem is determining which orders to include in the calculation of this indicator and when to count them. Not every order placed is paid for, and not every order placed and paid for is shipped to the buyer. Not every shipment is picked by the buyer, and some of the goods purchased may be returned in accordance with applicable law. All of these aspects make it necessary to clarify what exactly is analyzed and which orders are taken into account when calculating the AOV value. The simplest way (because it does not require additional integration with the warehouse system) is to count the indicator for orders placed, but the analyst must be aware of possible discrepancies with results that only take into account goods sold (including, in particular, the returns).

All of the above metrics can be used to measure the effectiveness of dedicated user interfaces. However, it should be noted that the CR and AOV metrics measure macroconversions, whereas the CTR metric measures microconversions. Undoubtedly, increasing conversions at the macro level is key, as it translates directly into increased e-shop revenue. On the other hand, focusing only on macroconversion limits the analysis of the impact of the user interface on those customers who ultimately decide to complete the purchase. This approach makes it impossible to assess the impact of UI changes on the behavior of customers who, for whatever reason, did not place an order. In contrast, microconversion analysis using CTR can be too granular. It allows you to see how a UI change affects a single customer behavior (e.g., clicking on a banner with an offer) but does not provide any insight into the related behaviors that make up the customer journey in an online store. The solution may be to introduce a metric with a level of detail that lies between macroconversion and microconversion —the **partial conversion rate (PCR)** [23].

The purpose of the PCR is to assess the extent to which the customer's behavior aligns with the expected use of the online shop. The available functions and the user interface determine the actions a customer can take in a specific area of the e-shop. For example, a customer who is on a product card can:

• Add the product to the cart
• Add product to favorites

- Select a product attribute
- Collapse/expand product description
- Enlarge photo from the gallery
- Return to home page
- Go to the category listing

and much, much more.

An analogous list of available actions can be prepared for each page in the e-shop. Undoubtedly, from an entrepreneur's perspective, some actions are desirable, while others are not. Additionally, some of the desirable actions may carry more significance than others. Therefore, it can be assumed that there is a set of actions expected by the e-shop owner. Identifying the locations in the shop where these actions can be performed and assigning importance (weighting) (Table 4.3) to them allow the design of the *expected customer journey*.

This information can also be presented in the form of a directed graph, where the nodes represent the expected states of the e-commerce system associated with a particular site, and the edges symbolize the customer's actions—the transitions between the states of the system—with weights assigned to them (Fig. 4.7).

The expected use of the e-shop can be compared with the actual behavior of the customer, rewarding actions in line with expectations with points (or even

Table 4.3 An example of the of expected actions

E-commerce page	Activity	Weight
Main page	Go to the *Category listing*	10
Main page	Go to the *Product page*	15
Category listing	Go to the *Product page*	10
Category listing	Go to another *Category listing*	5
Product page	Add to the cart	20
Product page	Open the gallery	5

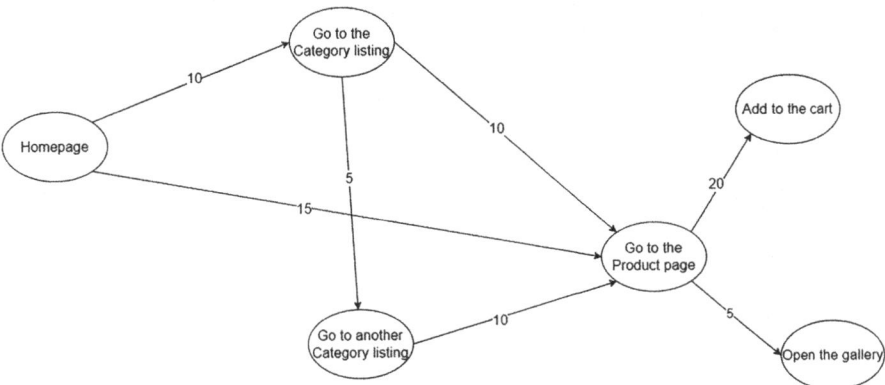

Fig. 4.7 An example of the of expected actions as a directed graph

deducting points to penalize behavior contrary to expectations). The PCR could then be calculated using the following formula:

$$PCR_c = \frac{1}{n} \sum_{i=1}^{n} \sum_{j=1}^{s} CVV_{ij} \qquad (4.4)$$

n is the number of sessions related to customers from the cluster c.
s is the number of activities within the session n.
PCR_c is the calculated PCR metric value for the cluster c.
CVV_{ij} is the score of an activity j during a session i.

The proposed PCR indicator offers a flexible way to verify the impact of interface changes on customer behavior. The expected customer journey can be of any length, avoiding the disadvantage of a CTR based on a single event. On the other hand, PCR does not necessitate the customer to make a purchase, allowing the consideration of the behavior of all e-shop customers.

The CR, AVR, and PCR metrics can be employed to test the effectiveness of user interface modifications for specific user groups (clusters). They are applicable when interface variants are designed by UX experts but can also serve as a basis for decision-making in a user interface auto-adaptation solution.

4.4 Auto-adaptation of User Interface

To achieve a satisfactory effect from the implementation of multivariant e-commerce user interfaces, the offered interface variants must align with the behavior of the customer groups identified during clustering. The decision on the appearance of the interface variants remains a fundamental issue. The most apparent solution is to entrust the task of tailoring interface variants to UX specialists. However, an alternative approach is implementing a mechanism for the automatic adaptation of interface variants.

The idea of such a tool is based on *microchanges* made sequentially to a variant of the interface served to a specific group of customers and verifying the impact of these microchanges with selected UI performance measures.

A microchange in this case should be understood as a single interface change or a set of related changes within a single area.

The general model of the auto-adaptation mechanism (Fig. 4.8) includes an initialization part and the component responsible for adding and evaluating microchanges. The initialization occurs for all e-commerce customers, while the interface customization (microchanges evaluation) is done in parallel and independently for each generated customer cluster. This means that the pace of change for each interface variant (assigned to the corresponding clusters) can be different.

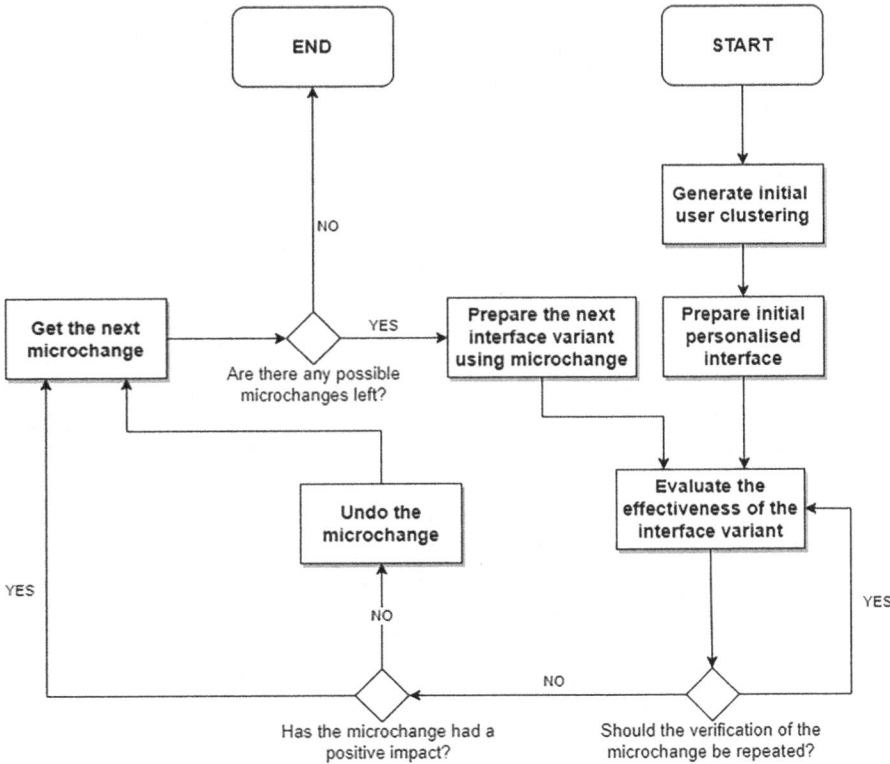

Fig. 4.8 The general model of the auto-adaptation mechanism

Initial Clustering

The first component of the solution is the preparation of user groups using the chosen clustering method. The choice of conditions for generating clusters is a key decision at the initialization stage. It is important to consider typical clustering quality indicators and the business requirements in terms of the number of interface variants that will be made available to customers. Examples of metrics that can be used to assess clustering quality include Silhouette score (SI), Calinski–Harabasz score (CH), and Davies–Bouldin score (DB).

The *Silhouette score (SI)*, introduced by Peter Rousseeuw [5], facilitates the interpretation and validation of consistency across data clusters. This metric estimates the similarity of an object (e.g., an e-shop customer) to objects in its cluster (cohesion) compared to other clusters (separation). The indicator ranges from -1 to +1, with higher values indicating a better fit of the object to its cluster and a worse fit to neighboring clusters.

The *Davies–Bouldin score (DB)*, proposed by David L. Davies and Donald W. Bouldin [6], measures the average similarity between each cluster and its most similar cluster. The similarity measure is the distance between cluster centroids and the dispersion within clusters. The interpretation of this measure is the opposite of the Silhouette score—low values (tending toward 0) indicate good clustering, as it means that the clusters are well separated and distinct. Conversely, high values may indicate that the clusters are not well separated. To enhance result interpretation consistency, an *Inverse DB (IDB)* (Eq. 4.5) measure can be introduced, with high values indicating good clustering and low values indicating poor clustering:

$$\text{IDB} = \frac{1}{\text{DB}} \tag{4.5}$$

The *Calinski-Harabasz Index (CH)*, introduced by Tadeusz Calinski and Jerzy Harabasz [7], measures the ratio of between-cluster variance to within-cluster variance and aids in determining the optimal number of clusters in a dataset. High values of the indicator indicate well-separated clusters with low variance within each cluster, while low values suggest poorly separated clusters or an inappropriate number of clusters.

The values of each quality indicator (SI, CH, and IDB) can be calculated for each clustering performed. To select the optimal number of clusters within the method, the results obtained can be standardized if necessary (taking the best result within a metric as 100% and calculating values for the worst results proportionally). Then, the values can be weighted according to the priorities adopted.

However, the ability to develop and maintain a dedicated interface depends on the size of the group for which it would be implemented. The division of e-commerce customers into multiple groups significantly impacts the size of clusters. For clusters with a small number of users, the time required to adapt the interface to them can be very long, and the business benefits are questionable.

The measures used to assess the alignment of clustering results with business requirements can be:

- The size of the smallest cluster in relation to the size of the entire analyzed population of customers
- Cluster spread index (CSI) calculated according to the formula:

$$\text{CSI} = \frac{\text{CS}_{\max} - \text{CS}_{\min}}{\text{CM}_{\max}} \tag{4.6}$$

where $\text{CS}_{\max}$ is the size of the most numerous cluster and $\text{CS}_{\min}$ is the size of the least numerous cluster.

- Standard deviation (SD) of cluster sizes

In the case of the CSI and SD indicators, the combinations [clustering method—number of clusters] for which the values were lowest should be considered the best. The higher the values these indicators take, the more the clustering effects deviate from the expectations arising from the business context.

The initial clustering parameter selection can be described by the following steps [10]:

1. Identification of clustering methods and parameters to be validated for potential use to generate initial clusters.
2. Clustering using a specific method and parameters.
3. Calculation of clusters and clustering quality metrics (SI, CH, IDB) for the analyzed method and its parameters.
4. Changing clustering parameters if all selected options have not yet been analyzed.
5. Clustering duration analysis and potential removal of methods too computationally complex.
6. Rejection of clustering options that lead to clusters that do not meet business requirements for the number of customers in the clusters.
7. Optional standardization of clustering quality metrics values (sSI, sCH, and sIDB) (Eq. 4.7).

$$\text{COE} = \frac{W_s * \text{sSI} + W_c * \text{sCH} + W_d * \text{sIDB}}{W_s + W_c + W_d} \qquad (4.7)$$

where COE is the number that determines the evaluation of the clustering option and W_x is the weight assigned to the clustering quality metric.

8. Select the method and the number of clusters that offer the best clustering quality indicators and best meet business requirements.

The calculated COE values for each clustering option can either provide a basis for recommending the choice of clustering conditions if the decision is to be made by a human or decide the choice of clustering conditions if the decision is to be made automatically.

The choice of the clustering method is crucial for the operation, and it is beneficial if it can be made by an expert who considers the results of the analysis of the available options and selects the best one, treating the system recommendation as support. However, in a situation where automatic selection is necessary (the solution delivered as a product in the SaaS model), the presented algorithm allows a fairly objective evaluation of the possible clustering options and the selection of the one with the highest evaluation.

Microchange Acceptance

The automatic adaptation of the user interface is based on the implementation of microchanges and the verification of their impact on e-commerce performance. Therefore, before starting the automatic adaptation of interface variants, it is necessary to create a list of possible microchanges and specify conditions for accepting or rejecting the implemented microchange. The list of all possible microchanges forms the basis for making changes to the interface variants. Some microchanges may be interdependent and should be implemented simultaneously. On the other hand, some microchanges may be mutually exclusive. This means that

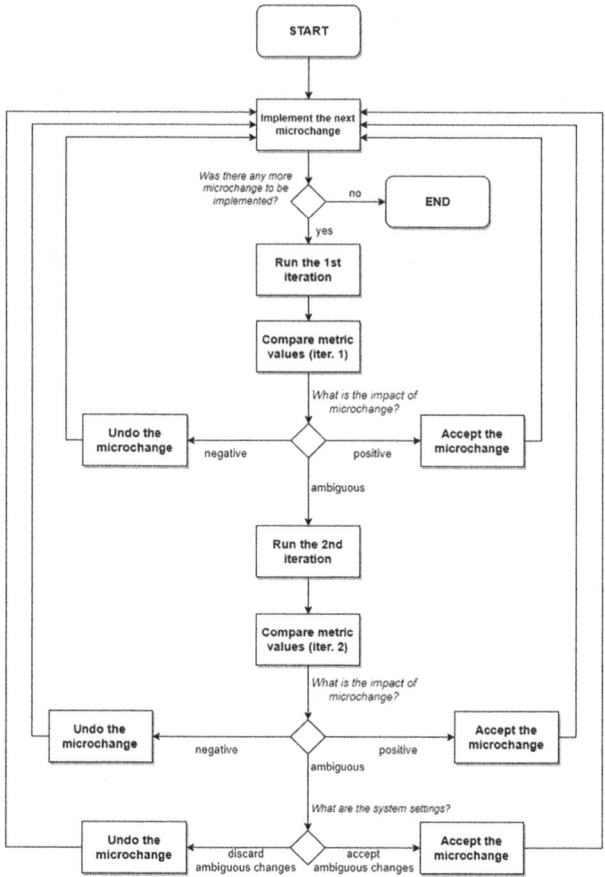

Fig. 4.9 Microchange acceptance algorithm

once a microchange is accepted, all potential microchanges that exclude it should automatically be rejected.

The proposed microchange acceptance algorithm (Fig. 4.9) is based on three e-commerce performance metrics, CR, AOV, and PCR, but can be extended with other metrics.

The analysis of the impact of the implemented microchange involves two stages. In the first stage, data is collected on the behavior of users who are randomly served either a dedicated (with implemented microchanges) or a standard interface variant. The study duration may be limited by:

- Time (e.g., in days or weeks)
- The minimum number of sessions during which the studied interface versions were shown
- The minimum number of orders placed using the tested UI variants

In the second phase, the values of the PCR, AOV, and CR indices are calculated for both the study group and the comparison group.

The acceptance of a microchange in the first iteration of verification can be conditional on the simultaneous fulfillment of three conditions:

1. $CR(d) > CR(s)$
2. $AOV(d) > AOV(s)$
3. $PCR(d) > PCR(s)$

where d represents the values obtained for the dedicated interface and s the indicator values for the standard interface.

The requirement for rejecting the microchange is similar—the simultaneous fulfillment of three conditions:

1. $CR(d) < CR(s)$
2. $AOV(d) < AOV(s)$
3. $PCR(d) < PCR(s)$

In other cases, if the first review does not result in a decision, the system proceeds to a second iteration.

The second analysis is similar to the first. The data is collected in the same way, but the difference is in the way the values of the PCR, AOV, and CR indicators are calculated. This time, the calculations can take into account the client's behavior during the first and second iterations, incorporating twice as much data as before. The requirements for accepting or rejecting a microchange may also differ. In this case, only two of the three acceptance conditions mentioned earlier need to be met for a change to be confirmed. Similarly, two of the three rejection conditions should be fulfilled for a microchange to be withdrawn.

To avoid decisions based on minor differences in indicator values, a minimum acceptable difference (e.g., 5%) can be specified. If the measured difference is less than the threshold value during the experiment, the algorithm will not consider it significant and will not make a decision based on it. The use of an acceptance threshold can also lead to a modification of the criteria for confirming or withdrawing a microchange. For example, it can be assumed that if a modification positively impacts at least one performance indicator (the difference in favor of the dedicated interface exceeds the threshold value) while not negatively impacting others (any differences to the detriment of the dedicated interface are below the threshold value), then such a microchange should be accepted. Similarly, a modification can be rejected if it strongly negatively affects at least one of the indicators while having little (below the threshold) positive impact on the other indicators.

If there is still no decision to accept or reject the microchange after the second iteration, the change is either accepted or rejected depending on the tool settings. If the auto-adapt mechanism is overseen by an administrator, there is an additional option to leave the decision to the administrator.

Once the analyzed verification is complete, the next microchange from the list is implemented and evaluated. The mechanism continues until the last predefined microchange has been verified. When all the predefined microchanges have been analyzed, the interface variant for a particular cluster of e-commerce customers enters the stabilization phase. However, if new ideas for microchanges emerge, they can be added to the list, and the mechanism returns to the phase of evaluating their impact on the effectiveness of the interface variant.

The described approach forms the basis for research related to automating the implementation of modifications within a multivariant e-commerce user interface. The implementation, validation, and verification of this approach are beyond the scope of this publication but can be carried out during future research.

References

1. Rahutomo R, Lie Y, Perbangsa AS, Pardamean B (2020) Improving conversion rates for fashion e-commerce with A/B testing. In: Conference: 2020 international conference on information management and technology: virtual conference. https://doi.org/10.1109/ICIMTech50083.2020.9210947
2. Kruger RM, Gelderblom H, Beukes W (2016) The value of comparative usability and UX evaluation for e-commerce organisations. In: Conference: CONF-IRM 2016 Proceedings. Paper 9
3. Syafrizal FA, Heroza RI, Ermatita, Firdaus MA, Putra P, Atrinawati LH, Adrian M (2016) Using ISO 9241-11 to identify how e-commerce companies applied UX guidelines. Inform Jurnal Ilmiah Bidang Teknologi Informasi dan Komunikasi. https://doi.org/10.25139/inform.v7i1.4443
4. Sabanoglu T (2022) Metrics used by U.S. retailers to measure personalization initiative success 2019. Statista. https://www.statista.com/statistics/1115435/metrics-used-by-us-retailers-to-measure-personalization-initiative-success/. Cited 24 Sep 2023
5. Rousseeuw PJ (1987) Silhouettes: a graphical aid to the interpretation and validation of cluster analysis. J Comput Appl Math. https://doi.org/10.1016/0377-0427(87)90125-7
6. Davies DL, Bouldin DW (1979) A cluster separation measure. IEEE Trans Pattern Anal Mach Intell. https://doi.org/10.1109/TPAMI.1979.4766909
7. Calinski T, Harabasz J (1974) A dendrite method for cluster analysis. Commun in Stat. https://doi.org/10.1080/03610927408827101
8. Burstein D (2012) Marketing research chart: average website conversion rates, by industry. MarketingSherpa. https://www.marketingsherpa.com/article/chart/average-website-conversion-rates-by. Cited 24 Sep 2023
9. Nielsen J (2013) Conversion rates. Nielsen Norman Group. https://www.nngroup.com/articles/conversion-rates/. Cited 24 Sep 2023
10. Wasilewski A (2024) Customer segmentation in e-commerce: a context-aware quality framework for comparing clustering algorithms. J Internet Services Appl 15:160–178

Chapter 5
Evaluation of a Multivariant Interface Implementation

5.1 Introduction to Research

The personalization of the user interface layout, using machine learning to group customers to serve them dedicated UI variants, seems to be an intriguing direction for e-commerce development. In theory, such a solution should increase customer satisfaction and, as a result, positively influence purchase decisions, thereby increasing the business benefits for the e-commerce owner. Whether such theoretical assumptions are reflected in practice remains an open question. The key, therefore, is to verify the practical feasibility of collecting information on e-commerce customer behavior, processing it in a useful way for UI personalization, and designing and delivering dedicated interface variants. In addition, such verification should include an analysis of the performance of the personalized UI—its impact on key indicators describing the effectiveness of the e-commerce business.

The practical verification of a comprehensive solution to realize the concept of a multivariant e-commerce user interface is not a simple matter. It requires the implementation of a system that combines modules responsible for collecting and processing customer behavior data and for designing, delivering, and monitoring dedicated interface variants. Such a software platform (called AIM^2) has been designed and implemented by Fast White Cat S.A. within the project *"Self-adaptation of the online store interface for the customer requirements and behavior"* cofinanced by the NCBR (National Centre for Research and Development) under Sub-Action 1.1.1 of the Operational Programme Intelligent Development 2014–2020. This complex solution allows for verifying the effectiveness of using personalized UIs in e-commerce.

The AIM^2 platform was initially designed to interact with e-shops based on the Adobe Magento solution and uses *Progressive Web Application* (PWA) technology to provide the user interface. The final solution is open and can be applied to e-

A. Wasilewski, *Multi-variant User Interfaces in E-commerce*, Progress in IS, https://doi.org/10.1007/978-3-031-67758-8_5

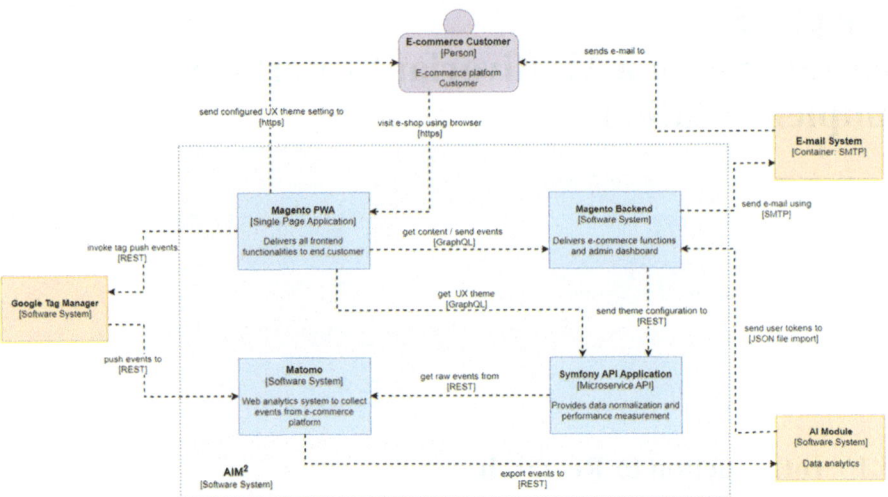

Fig. 5.1 The general architecture of the AIM^2 platform

shops using different software, in accordance with the Service Oriented Architecture paradigm [1].

The general architecture of the AIM^2 platform is shown in Fig. 5.1.

The platform has been developed using the following technologies:

- PHP 8.1+: the latest version of the PHP language, providing the best performance and support for object-oriented software development in hexagonal architecture.
- Symfony 6.1+: the latest version of the Symfony framework supporting component-based programming and thus the reuse of previously created code, especially in the application and integration layers.
- API Platform 3+: the latest version of the library accelerating the creation of API services, enabling rapid development of applications (RADs) compliant with API standards (openapi 3, graphql). In the latest version, it also supports architectures based on ports and adapters (Providers and Processors).
- Docker: an application containerization environment that enables a unified technology stack from software development to maintenance in production.

The Google Tag Manager and Matomo (version 4.13.0) tools are responsible for collecting customer behavior data. The information collected is fed into the module responsible for grouping customers based on ML algorithms (AI module) and is retrieved by the platform management application (Symfony API application) to generate customer behavior reports.

The AI/ML module has functions for clustering customers using k-means, k-medians, BIRCH, agglomerative, spectral, and DBSCAN algorithms. The clustering results, including the characteristics of the clusters and the customers assigned to the clusters, are passed to the Symfony API application.

The Symfony API application also includes the configuration of user interface variants (themes) and the assignment of these variants to specific customers. This information is retrieved by the front end when the customer opens the first page of the e-shop.

On the e-shop side (backend and frontend), interface modifications are implemented that can be combined into variants dedicated to specific groups (clusters) of customers. At the time of the study, the AIM2 platform implemented 26 modifications that could be used to personalize the interface. They allowed the modification of the basic elements of the e-shop—home page, listing, and product card. A list of implemented modifications is provided in Appendix B.

Data on the behavior and purchasing decisions of customers of OTCF's[1] online store https://4fstore.lv/ was used to test the concept of multivariant e-commerce interfaces in practice. The research consisted of two elements—the selection of the clustering method and the analysis of the effectiveness of the UI variants designed and delivered. The review lasted 7 months, during which time data was collected from 607570 user sessions.

5.2 Selection of an Optimal Clustering Method

5.2.1 Introduction

Serving multivariant user interfaces does not inherently require advanced methods of processing information about customer behavior. Customer grouping can be done using classical methods of segmentation in e-commerce. In this case, knowledge of users' final decisions (e.g., orders placed, requests sent, etc.) or their characteristics (e.g., age, gender, location, etc.) is usually used. However, nowadays it is becoming increasingly difficult to obtain complete data that allows for classic segmentation, due to the high importance that customers attach to the security of their data.

In such a situation, the solution may be to use information that is collected anonymously and with privacy when customers visit the e-shop. However, this approach poses a different kind of challenge due to the large amount of data collected. Given that a typical e-commerce conversion rate is a few percent, there may be dozens of times more information about all customers than there is about the customers who ultimately place an order. The answer may lie in the use of clustering algorithms capable of processing large amounts of data and clustering customers based on it.

---

[1] OTCF specializes in the design, production, and sale of sports clothing and accessories for amateurs and professionals. The company's portfolio includes the 4F, 4F Junior, 4F Fuel, 4F Spot, and Outhorn brands, as well as the SportStyleStory multibrand sports stores. OTCF S.A. is also exclusive distributor of the Under Armour brand on the Polish market.

Goals

The goals of the research on the clustering methods were:

(G1.1): Analyzing implemented clustering algorithms for suitability in clustering customers for serving multivariant UIs.

(G1.2): Verification of the method for selecting the optimal clustering algorithm.

(G1.3): Selection of the best (or best) clustering algorithm(s) for further research on the effectiveness of multivariate UIs.

Research Questions

To achieve the stated objectives of the study, the following research questions were addressed:

(Q1.1): What criteria should determine the choice of a clustering algorithm for serving dedicated UIs?

(Q1.2): How should different clustering methods be analyzed before selecting an algorithm for multivariant UIs?

(Q1.3): Which results obtained should exclude the clustering method from further analysis?

(Q1.4): Which clustering method should be selected for further research on the effectiveness of multivariant UIs?

Research Methodology

The research conducted was an experiment using data collected from a real e-shop. This involved collecting data on user behavior and then grouping customers using selected clustering methods. The analysis included six methods representing different clustering approaches: K-means, K-medians, BIRCH, DBSCAN, spectral, and agglomerative.

For each method (except DBSCAN), clustering was performed assuming several clusters from 3 to 8. The number of clusters was determined by business requirements for the number of customers in the smallest cluster. It was assumed that there would be no economic justification for investing in a dedicated interface variant if less than 5% of the customer population were in the cluster. Preliminary studies have shown that in practice, with more than eight clusters, the smallest clusters are well below this threshold.

For each combination of the clustering method and the number of clusters, information was collected on the clustering time (without preprocessing), the values of the clustering quality indicators, the size of the clusters, and the main characteristics of the clusters (average number of shares, events, and revenues).

The evaluation of clustering methods was carried out along three dimensions:

1. Technical requirements
2. Clustering quality indicators
3. Specific business requirements

The *technical requirements* analysis includes the resources required (including CPU and memory) and the clustering time, which is particularly important for large datasets and frequent reclustering to update customer groups in clusters. The importance of this factor is relatively low but may increase if clustering takes longer than the set reclustering cycle or if infrastructure costs are time-dependent (e.g., time charges for the use of processors, memory, or other resources).

The measure of this aspect can be T_c—the time it takes to perform clustering. It was determined for each combination of [clustering method—number of clusters], except for the DBSCAN method, for which only one value of T_c was determined. Another way to measure this dimension can be the extent to which resources (CPU and memory) are consumed when clustering a given set of input data.

Clustering quality indicators allow different aspects of clustering quality and validity to be assessed and are applicable regardless of the business context. They belong to the group of model evaluation-based methods. The research carried out considered well-known indicators such as the Silhouette Score, Calinski–Harabasz Index, and Davies–Bouldin Score.

To standardize the indicator values, it was assumed that the weights (W_s, W_c, W_d) of each of the measures considered (SI, CH, IDB) would be the same at 33.3(3)%.

Specific business requirements arise from the business context of clustering and the reasons for clustering. They determine the practical suitability of clusters for business use. For multivariant user interfaces, requirements will focus on the distribution of clients in clusters and the minimum number of clients in a cluster.

The research used collected data on user sessions, which included the following information (Fig. 5.2):

- **Session**—user session's fingerprint
- **User ID**—customer's UUID
- **Type**—user activity type (e.g., event, listing, product, homepage, checkout, blog, other)
- **Category**—activity category for events (e.g., product page, cart, search, purchase)
- **Action**—activity action for event (e.g., minicart, select size, select color, thumbnail click, click accordion, image click)

fingerprint	userId	type	category	action	name	url	time	visitKey
fe9fe562d186a3b9	01470a1d-e9fa-45fe-9db2-b7cf1b101231	event	ProductPage	selectSize	S	https://4fstore.lv/sieviesu-sporta-dzemperis-no-organiskas-kokvilnas-4fs23tswsf	28.04.2023 22:51	event-ProductPage-selectSize-S-sieviesu-sporta
fe9fe562d186a3b9	01470a1d-e9fa-45fe-9db2-b7cf1b101231	event	ProductPage	selectColor	Violeta	https://4fstore.lv/sieviesu-sporta-bikses-no-organiskas-kokvilnas-4fs23trof386-42	28.04.2023 22:52	event-ProductPage-selectColor-Violeta-sieviesu-sporta
fe9fe562d186a3b9	01470a1d-e9fa-45fe-9db2-b7cf1b101231	listing				https://4fstore.lv/meitenes/apgerbs/dzemperi/cena-9-20.html	28.04.2023 22:53	listing-meitenes-apgerbs-dzemperi-cena-9
fe9fe562d186a3b9	01470a1d-e9fa-45fe-9db2-b7cf1b101231	product				https://4fstore.lv/meitenu-sporta-dzemperis-hudijs-4fjaw2tswsf450-45n.html	28.04.2023 22:53	product-meitenu-sporta
fe9fe562d186a3b9	01470a1d-e9fa-45fe-9db2-b7cf1b101231	product				https://4fstore.lv/meitenu-sporta-dzemperis-hudijs-4fjaw2tswsf450-45n.html	28.04.2023 22:53	product-meitenu-sporta
fe9fe562d186a3b9	01470a1d-e9fa-45fe-9db2-b7cf1b101221	account				https://4fstore.lv/customer/account/	28.04.2023 22:54	account-customer-account
fe9fe562d186a3b9	01470a1d-e9fa-45fe-9db2-b7cf1b101231	listing				https://4fstore.lv/meitenes/apgerbs/bikses.html	28.04.2023 22:54	listing-meitenes-apgerbs-bikses
fe9fe562d186a3b9	01470a1d-e9fa-45fe-9db2-b7cf1b101231	listing				https://4fstore.lv/meitenes/apgerbs/bikses.html?sortAttribute=price&sortDirecti	28.04.2023 22:54	listing-meitenes-apgerbs-bikses
fe9fe562d186a3b9	01470a1d-e9fa-45fe-9db2-b7cf1b101231	event	ProductPage	thumbnailClick		https://4fstore.lv/joger-tipa-meitenu-mikroflisa-bikses-4fjaw21trof016-52s.html	28.04.2023 22:55	event-ProductPage-thumbnailClick-joger-tipa
fe9fe562d186a3b9	01470a1d-e9fa-45fe-9db2-b7cf1b101231	event	ProductPage	thumbnailClick		https://4fstore.lv/joger-tipa-meitenu-mikroflisa-bikses-4fjaw21trof016-52s.html	28.04.2023 22:55	event-ProductPage-thumbnailClick-joger-tipa
fe9fe562d186a3b9	01470a1d-e9fa-45fe-9db2-b7cf1b101231	product				https://4fstore.lv/joger-tipa-meitenu-sporta-bikses-4fjaw2trof279-5n.html	28.04.2023 22:55	product-joger-tipa
fe9fe562d186a3b9	01470a1d-e9fa-45fe-9db2-b7cf1b101231	event	ProductPage	thumbnailClick		https://4fstore.lv/joger-tipa-meitenu-sporta-bikses-4fjaw2trof279-5n.html	28.04.2023 22:55	event-ProductPage-thumbnailClick-joger-tipa
a7528b9e395c141d	00202b4c-ddfd-4808-b8cb-c9652fc2f087	product				https://4fstore.lv/zenu-saulesbriles-4fjs23asunm024-21s.html?utm_content=carsl	29.04.2023 11:12	product-zenu-saulesbriles
a7528b9e395c141d	00202b4c-ddfd-4808-b8cb-c9652fc2f087	product				https://4fstore.lv/zenu-saulesbriles-4fjs23asunm024-21s.html	29.04.2023 11:13	product-zenu-saulesbriles
8e7c1000321a78fa	0026ed6c-3761-40c9-a813-6c1e0671ae64	event	ProductPage	selectSize	S	https://4fstore.lv/sieviesu-trekinga-virsjaka-membrana-10-0-4fs23tjaf251-2s.html	30.04.2023 09:35	event-ProductPage-selectSize-S-sieviesu-trekinga
8e7c1000321a78fa	0026ed6c-3761-40c9-a813-6c1e0671ae64	checkout				https://4fstore.lv/checkout-cart.html	30.04.2023 09:35	checkout-checkout-cart
8e7c1000321a78fa	0026ed6c-3761-40c9-a813-6c1e0671ae64	listing				https://4fstore.lv/izpardosana/sievietes/apgerbs.html	30.04.2023 09:49	listing-izpardosana-sievietes-apgerbs
8e7c1000321a78fa	0026ed6c-3761-40c9-a813-6c1e0671ae64	event	ProductPage	selectColor	Violeta	https://4fstore.lv/sieviesu-pilsetas-virsjaka-membrana-5-0-4fs23tjaf232-54s.html	30.04.2023 09:50	event-ProductPage-selectColor-Violeta-sieviesu-pilsetas
8e7c1000321a78fa	0026ed6c-3761-40c9-a813-6c1e0671ae64	product				https://4fstore.lv/sieviesu-pilsetas-virsjaka-membrana-5-0-4fs23tjaf232-52s.html	30.04.2023 09:50	product-sieviesu-pilsetas
8e7c1000321a78fa	0026ed6c-3761-40c9-a813-6c1e0671ae64	event	ProductPage	selectColor	Peleka	https://4fstore.lv/sieviesu-pilsetas-virsjaka-membrana-5-0-4fs23tjaf232-52s.html	30.04.2023 09:50	event-ProductPage-selectColor-Peleka-sieviesu-pilsetas
8e7c1000321a78fa	0026ed6c-3761-40c9-a813-6c1e0671ae64	listing				https://4fstore.lv/izpardosana/sievietes/apgerbs/page-2.html	30.04.2023 09:51	listing-izpardosana-sievietes-apgerbs-page-2
8e7c1000321a78fa	0026ed6c-3761-40c9-a813-6c1e0671ae64	listing				https://4fstore.lv/izpardosana/sievietes/apgerbs/page-5.html	30.04.2023 09:53	listing-izpardosana-sievietes-apgerbs-page-5
8e7c1000321a78fa	0026ed6c-3761-40c9-a813-6c1e0671ae64	product				https://4fstore.lv/sieviesu-trekinga-virsjaka-membrana-10-0-4fs23tjaf251-2s.html	30.04.2023 09:53	product-sieviesu-trekinga
8e7c1000321a78fa	0026ed6c-3761-40c9-a813-6c1e0671ae64	checkout				https://4fstore.lv/checkout-cart.html	30.04.2023 09:54	checkout-checkout-cart
7da9b5703c6b3e19	01470a1d-e9fa-45fe-9db2-b7cf1b101231	product				https://4fstore.lv/sieviesu-flisa-dzemperis-ar-stavapkakli-regular-4faw2uflef015-	30.04.2023 13:25	product-sieviesu-flisa
7da9b5703c6b3e19	01470a1d-e9fa-45fe-9db2-b7cf1b101231	event	Cart	miniCart	Open	https://4fstore.lv/sieviesu-flisa-dzemperis-ar-stavapkakli-regular-4faw2uflef015-	30.04.2023 13:25	event-Cart-miniCart-Open
7da9b5703c6b3e19	01470a1d-e9fa-45fe-9db2-b7cf1b101231	listing				https://4fstore.lv/meitenes/apgerbs/bikses.html	30.04.2023 13:25	listing-meitenes-apgerbs-bikses
7da9b5703c6b3e19	01470a1d-e9fa-45fe-9db2-b7cf1b101231	listing				https://4fstore.lv/sievietes/apgerbs/bikses.html	30.04.2023 13:26	listing-sievietes-apgerbs-bikses
7da9b5703c6b3e19	01470a1d-e9fa-45fe-9db2-b7cf1b101231	listing				https://4fstore.lv/sievietes/apgerbs/bikses/page-2.html	30.04.2023 13:27	listing-sievietes-apgerbs-bikses-page-2
7da9b5703c6b3e19	01470a1d-e9fa-45fe-9db2-b7cf1b101231	product				https://4fstore.lv/sieviesu-sporta-dzemperis-no-organiskas-kokvilnas-4fs23tswsf	30.04.2023 13:27	product-sieviesu-sporta
7da9b5703c6b3e19	01470a1d-e9fa-45fe-9db2-b7cf1b101231	event	Cart	miniCart	Open	https://4fstore.lv/sieviesu-sporta-dzemperis-no-organiskas-kokvilnas-4fs23tswsf	30.04.2023 13:27	event-Cart-miniCart-Open
7da9b5703c6b3e19	01470a1d-e9fa-45fe-9db2-b7cf1b101231	product				https://4fstore.lv/sieviesu-sporta-bikses-no-organiskas-kokvilnas-4fs23trof386-42	30.04.2023 13:27	product-sieviesu-sporta
260288cddecf74f5	003697ff-aa94-4fa9-920c-517042c05abe	product				https://4fstore.lv/bezsuvju-sieviesu-longsleve-crop-top-4faw2tlonf073-20s.html?	30.04.2023 15:26	product-bezsuvju-sieviesu
260288cddecf74f5	003697ff-aa94-4fa9-920c-517042c05abe	product				https://4fstore.lv/sieviesu-skriesanas-legingi-ar-augstu-vidukli-4faw2tftrf123-20s.	30.04.2023 15:26	product-sieviesu-skriesanas
f37532ac13d8a86b	00789d89-e6e9-4b6c-9069-89841d80889d	product				https://4fstore.lv/sieviesu-joger-tipa-sporta-bikses-izgatavotas-no-parstradatiem	30.04.2023 20:00	product-sieviesu-joger
7da9b5703c6b3e19	01470a1d-e9fa-45fe-9db2-b7cf1b101231	product				https://4fstore.lv/sieviesu-sporta-dzemperis-no-organiskas-kokvilnas-4fs23tswsf	30.04.2023 21:05	product-sieviesu-sporta
742d22ffb0de52de	0070f235-332b-4341-8044-924a27efc385	product				https://4fstore.lv/sieviesu-pilsetas-metelis-membrana-5-0-4fs23tjaf369-20s.html	01.05.2023 03:43	product-sieviesu-pilsetas
742d22ffb0de52de	0070f235-332b-4341-8044-924a27efc385	listing				https://4fstore.lv/izpardosana/sievietes.html	01.05.2023 03:45	listing-izpardosana-sievietes

Fig. 5.2 Source data on user activities

- **Name**—additional description of the action (e.g. open, close, size value, color value)
- **URL**—URL of the page on which the activity was performed
- **Time**—date and hour of the action

The above data is used to generate an activity identifier (**visitKey**), which is the basis for calculating reports for actions and action sequences.

The analysis, which included a predefined number of clusters ranging from 3 to 8, was carried out in the following stages:

1. For 18 days and 65174 sessions
2. For 36 days and 130969 sessions
3. For 54 days and 197363 sessions
4. For 72 days and 261774 sessions
5. For 90 days and 310607 sessions

The main objective of the first four iterations was to determine the best clustering method (or methods) for testing the effectiveness of multivariate user interfaces, while the main objective of the last iteration was to deepen the analysis of the effect of input set size on clustering time.

The results of the fourth iteration of the study formed the basis for selecting a clustering method to test the effectiveness of the dedicated interface variants.

5.2.2 Results

Clustering Requirements

The time required to determine the clusters is the sum of the preprocessing time and the clustering time. The first component is independent of the chosen clustering method, but the second depends directly on the computational complexity of the algorithm. In addition to the algorithm itself, the clustering time is also affected by the size of the dataset. The tests were carried out on a server with 70 GB of RAM and 16 processor threads.

To verify the effect of the clustering algorithm and the size of the dataset on the time needed to compute the clusters, clustering was performed with the implemented methods on five different datasets, obtained by increasing the date range of the analyzed customer behavior.

The results obtained are shown in Table 5.1. The size of the dataset was measured in terms of the number of user sessions analyzed, and the time required for clustering was measured in seconds. For methods that require a predefined number of clusters, the algorithms were assumed to generate six clusters containing e-commerce customer IDs.

The size of the datasets changed almost linearly as a result of adding very similar periods (18 days) to successive iterations of the study. In contrast, the time required for clustering grew faster than the size of the processed dataset for most of the

Table 5.1 Clustering duration (in seconds)

Dataset size	Agglom.	BIRCH	DBSCAN	K-means	K-medians	Spectral
65174	180	128	257	130	166	202
130969	198	152	268	158	195	214
197363	445	304	502	226	338	612
261774	856	449	856	375	675	1203
310607	1399	530	1112	459	782	1992

algorithms analyzed. The results showed that the fastest clustering method was the K-means algorithm, although even in this case, the time required to generate customer groups grew faster than the size of the input dataset. This means that as the amount of training data increases, the time required to generate clusters can be expected to increase significantly, which can have a negative impact on the performance of the overall solution. It is undoubtedly possible to scale up the clustering mechanism by adding resources such as processors and memory, but this involves an increase in maintenance costs, which is not insignificant when analyzing the economics of the system.

For very large datasets, the criterion of clustering time and resource consumption (Fig. 5.3) may be crucial in the choice of clustering method and in particular may exclude the most computationally intensive algorithms (e.g., agglomerative and spectral clustering) from the available options. In such a situation, methods that can handle large amounts of input data, such as K-means, will be of particular interest.

Clustering Quality Indicators

The values of selected clustering quality indicators calculated during the first iteration of the study are given in Table 5.2. The same table shows the values of the COE index, which is a synthetic measure of the standardized values of each detailed clustering quality measure.

The highest values for each number of clusters analyzed are highlighted in bold, while the lowest values are highlighted in italics.

The results allow initial conclusions to be drawn about the quality of the clusters obtained:

- For the *Silhouette score*:

  - The best results were obtained with the K-means (four times) and BIRCH (two times) methods, while the worst results were obtained with the spectral (five times) and agglomerative (one time) method.
  - Values were characterized by relatively low variation and ranged from 0.1603 to 0.2868.

- For the *Calinski–Harabasz score*:

  - The best results always appeared for the K-means method, while the worst results appeared only for the spectral method.

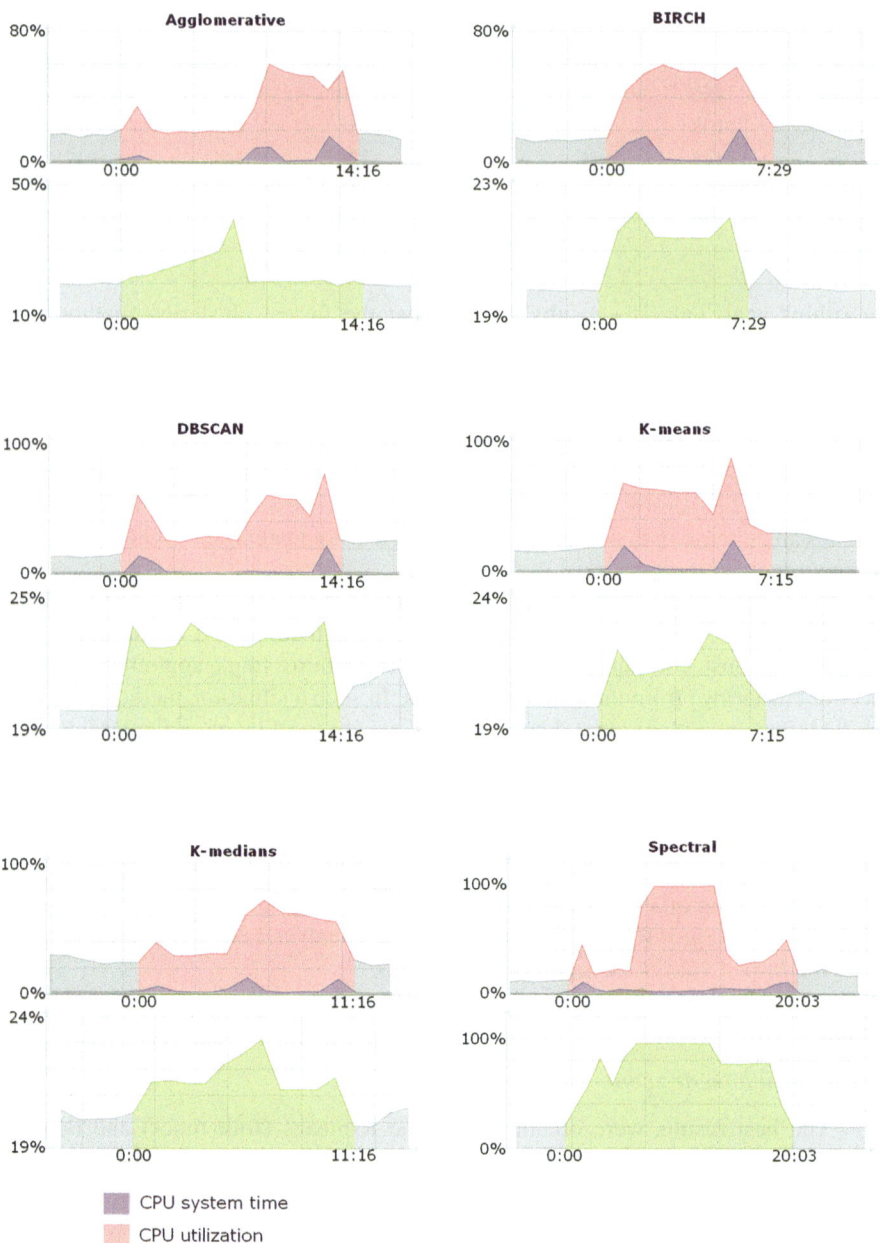

Fig. 5.3 Resource usage with different clustering methods (261774 user sessions)

Table 5.2 Quality metrics for the dataset of 130969 sessions

Clusters	Algorithm	Silhouette	Calinski–Harabasz score	IDB score	COE
3	Agglom.	*0.2212*	3091.1191	0.6052	**88.4826%**
	BIRCH	0.2769	2967.4840	0.3896	82.9134%
	K-means	**0.2856**	**3158.1121**	0.4211	87.5072%
	K-medians	0.2684	2596.7690	*0.3702*	77.0448%
	Spectral	0.2371	*1740.4476*	**0.6104**	*76.3359%*
4	Agglom.	0.2474	2739.2434	0.4463	79.9064%
	BIRCH	**0.2816**	2162.5672	*0.3665*	73.8090%
	K-means	0.2531	**2873.9028**	0.4666	**82.9937%**
	K-medians	0.2705	1711.4055	0,3744	68,1524%
	Spectral	*0.2346*	*1162.1868*	**0,5063**	*64.7479%*
5	Agglom.	0.2581	2305.3288	**0.4784**	**78.1648%**
	BIRCH	**0.2676**	1721.1137	0.3249	65.4547%
	K-means	0.2494	**2407.1629**	0.4318	75.9012%
	K-medians	0.2619	1201.7076	*0.3115*	58.6385%
	Spectral	*0.2395*	*1064.5904*	0.3416	*56.0957%*
6	Agglom.	0.2648	1993.3384	0.3786	70.6785%
	BIRCH	0.2563	1450.4368	0.3237	61.2222%
	K-means	**0.2761**	**2156.7161**	0.4486	**77.2057%**
	K-medians	0.2527	1402.4917	*0.2571*	*56.9855%*
	Spectral	*0.2238*	*840.8732*	**0.6691**	68.2195%
7	Agglom.	0.2717	1777.4750	0.3716	68.8509%
	BIRCH	0.2593	1261.6705	*0.3260*	59,6888%
	K-means	**0.2823**	**1911.2612**	0.3891	**72.3716%**
	K-medians	0.2603	871.4831	0.3272	*55.7519%*
	Spectral	*0.2308*	*790.6301*	**0.6538**	67.7360%
8	Agglom.	0.2693	1622.3826	0.3721	66.9696%
	BIRCH	0.2591	1134,9006	*0.3373*	58.8946%
	K-means	**0.2868**	**1738.9291**	0.3977	**71.4991%**
	K-medians	0.1953	854.3195	0.3493	*49.1195%*
	Spectral	*0.1603*	*129.8101*	**0.6260**	51.1893%
207	DBSCAN	0.2825	59.5577	0.7867	

- Values were characterized by a large variation and ranged from 129.8101 to
  3158.1121.

- For the *inverted Davies–Bouldin score*:

  - The best results were obtained with the spectral (five times) and agglomerative
    (once) methods, while the worst results were obtained with the BIRCH (three
    times) and K-median (three times) methods.
  - Values were characterized by medium variation and ranged from 0.2571 to
    0.6691.

- For the *COE score*:
  - The best results were obtained with the K-means (four times) and agglomerative (two times) methods, while the worst results were obtained with the spectral (three times) and K-median (three times) methods.
  - Values ranged from 49.1194% to 88.4826%.

Assuming that the COE index collectively assesses quality in terms of all the metrics considered, the best combinations can be tentatively identified [clustering method—number of clusters]:

1. Agglomerative, three clusters (88.48236%)
2. K-means, three clusters (87.9937%)
3. K-means, four clusters (82.9937%)
4. BIRCH, three clusters (82.9134%)
5. Agglomerative, four clusters (79.9064%)

Finally, it can be concluded that three methods deserve special attention during further verification of clustering results:

- **K-means method**, as it was the most frequent among the best results obtained, including the case of the synthetic indicator (COE)
- **Agglomeration method**, because of the double best performance for the COE indicator and the single best performance for the other indicators, although the single worst performance should also be noted
- **Spectral method**, as it appeared most often as the worst method but was also the best in terms of IDB rate

The other two methods (BIRCH and K-medians) did not stand out at this stage of the study. The DBSCAN method generated 207 customer clusters, which excluded it from detailed analysis, as it did not meet the initial assumptions on the number of clusters (maximum 8).

In the next iteration, the analysis was repeated for a doubled dataset, and the results are shown in Table 5.3.

The results obtained were similar to those obtained for a smaller dataset, indicating some reproducibility of the results.

This time, however, there was a clear advantage for the K-means method, which gave four times the highest *Silhouette score* values and six times for the *Calinski–Harabasz score* and *COE score* indices.

Again, it was found that the spectral method gave the best results for the IDB index, although when the results were aggregated to give the COE index, it was in last place four times.

Based on the COE index, it allows identifying the potentially best combinations [clustering method—number of clusters]:

1. K-means, three clusters (79.3987%)
2. Agglomerative, three clusters (75.3782%)
3. BIRCH, three clusters (74.7060%)

Table 5.3 Quality metrics for the dataset of 261774 sessions

Clusters	Algorithm	Silhouette	Calinski–Harabasz score	IDB score	COE
3	Agglom.	*0.1996*	9451.0481	0.6078	75.3782%
	BIRCH	0.2658	9150.8710	*0.3638*	74.7060%
	K-means	**0.2746**	**9847.0659**	0.4039	**79.3987%**
	K-medians	0.2624	9008.9582	0.3693	73.9883%
	Spectral	0.2482	*4.2057*	**1.0576**	*63.4786%*
4	Agglom.	*0.2256*	8478.2582	0.4522	70.3404%
	BIRCH	0.2300	6686.0545	0.3342	61.0814%
	K-means	0.2305	**8738.5335**	0.4514	**71.7935%**
	K-medians	0.2357	5107.5942	*0.3173*	*55.9008%*
	Spectral	**0.2436**	*3630.7851*	**0.7516**	65.5500%
5	Agglom.	0.2373	7136.1501	0.4826	68.1748%
	BIRCH	0.2338	5316.9126	0.3488	57.3673%
	K-means	0.2525	**7452.3560**	0.4820	**71.0716%**
	K-medians	**0.2537**	4034.6266	*0.2939*	*53.7202%*
	Spectral	*0.2227*	*2858.2959*	**0.7544**	60.4866%
6	Agglom.	*0.2212*	3091.1191	0.6052	61.7600%
	BIRCH	0.2395	4533.7015	0.3200	54.5065%
	K-means	**0.2535**	**6632.2424**	0.4388	**67.0536%**
	K-medians	0.2468	3937.8084	*0.2844*	*52.2527%*
	Spectral	0.2232	*2421.7450*	**0.7498**	58.9183%
7	Agglom.	0.2379	5481.4189	0.3658	58.9593%
	BIRCH	0.2380	3902.0445	0.3350	52.6688%
	K-means	**0.2591**	**5927.5637**	0.4392	**65.3628%**
	K-medians	0.2399	3595.0016	*0.2579*	49.4184%
	Spectral	*0.1582*	*127.7529*	**0.9079**	*48.2562%*
8	Agglom.	0.2318	4873.0594	0.3781	56.5479%
	BIRCH	0.2405	3456.2486	0.3247	51.1214%
	K-means	**0.2631**	**5311.3287**	0.3627	**61.3447%**
	K-medians	0.2439	2954.8645	*0.2618*	47.8563%
	Spectral	*0.1533*	*124.3091*	**0.8790**	*46.7369%*
645	DBSCAN	0.3148	76.5623	0.8010	

4. K-medians, three clusters (73.9883%)
5. K-means, four clusters (71.7935%)

The COE indicator favors a small cluster size, which in turn is not necessarily preferable for business reasons. If the options that generate three output clusters were removed, the list of best combinations [clustering method—number of clusters] would be as follows:

1. K-means, four clusters (71.7035%)
2. K-means, five clusters (71.0716%)

3. Agglomerative, four clusters (70.3404%)
4. Agglomerative, five clusters (68.1748%)
5. K-means, six clusters (67.0536%)

The results show that when the number of clusters is greater than 3, clustering using two methods—K-means and agglomerative—gives the best results. Therefore, it can be assumed that these two methods should be considered when choosing a clustering approach, taking into account clustering quality indicators. However, it should not be forgotten that in addition to assessing the quality of clustering, it is necessary to take into account the business requirements arising from the context of the clustering application. The final choice of clustering method and the number of clusters should take into account both factors—clustering quality measures and business requirements.

Specific Business Requirements
The preparation of interface variants for customer groups of an e-shop should be based on the ability to identify behavioral differences between groups and on an economic calculation. Assuming that a correctly chosen clustering method allows for a good differentiation of customer groups and that the analysis of actions and sequences of actions allows for the identification of key elements of the user interface that can influence users' decisions, the question of the rationality of designing dedicated UI variants remains.

Undoubtedly, it does not make sense to design an interface variant for a small group of customers, especially since in practice it is expected that only a part of them will visit the e-shop again. This means that the basic condition should be to determine the minimum number of users in the cluster for which it is worth designing a dedicated interface. This threshold can be defined absolutely, as a specific number of customers, or relatively, as a percentage of the total customer population. The second approach appears to be more flexible and scalable. For the study, it was assumed that the threshold for the minimum number of users in the cluster would be 5% of the total customer population. In Table 5.4, which summarizes the results of the first iteration of the analysis, the minimum cluster counts that do not exceed the accepted threshold are shown in italics.

In addition to the abundance of the smallest clusters, the results of the analysis were described through the abundance of the largest clusters and measures characterizing the distribution of cluster sizes—standard deviation (SD) and CSI. The lowest SD and CSI values for each number of result clusters are shown in bold, as for these indicators, the smaller the value, the better.

The main conclusions from the analysis of the results of the first iteration of the study are that:

- For the number of clusters from 3 to 5, the best results are obtained with agglomerative clustering, and for the number of clusters from 6 to 8, the best results are obtained with the K-means method.

Table 5.4 Cluster sizes (130969 sessions and 15330 customers)

Clusters	Algorithm	Biggest	Smallest	Std. dev.	CSI
3	Agglom.	7396	3474	**2040.1951**	**0.5303**
	BIRCH	7962	3109	2535.9521	0,6095
	K-means	7963	3164	2524.7944	0.6027
	K-medians	8967	2914	3350.9540	0.6750
	Spectral	11112	57	5588.2651	0.9949
4	Agglom.	4460	3199	**593.4254**	**0.2827**
	BIRCH	7962	660	3038.3933	0.9171
	K-means	4522	2843	819.8913	0.3713
	K-medians	9135	22	3926.4714	0.9976
	Spectral	11224	55	5267.2440	0.9955
5	Agglom.	4197	1769	**906.3868**	**0.5785**
	BIRCH	7962	660	2957.8552	0.9171
	K-means	4491	1428	1115.3011	0.6820
	K-medians	9721	267	4134.9286	0.9725
	Spectral	10641	57	4538.5270	0.9946
6	Agglom.	3474	1185	892.3383	0.6589
	BIRCH	7962	660	2810.6214	0.9171
	K-means	3474	1428	**738.4492**	**0.5889**
	K-medians	7961	540	2799.1029	0.9322
	Spectral	10687	7	4272.8072	0.9993
7	Agglom.	3474	656	1039.6887	0.8112
	BIRCH	7962	350	2688,6866	0.9560
	K-means	3474	1088	**760.4836**	**0.6868**
	K-medians	9366	87	3551.1520	0.9907
	Spectral	10365	7	3872.5520	0.9993
8	Agglom.	3474	656	1039.6887	0,8112
	BIRCH	7962	350	2550.01382	0.9560
	K-means	3473	832	**886.1080**	**0.7604**
	K-medians	6400	138	2728.7431	0.9784
	Spectral	14279	2	4999.8040	0.9999

- The lowest CSI values were obtained for the combination [clustering method—number of clusters]:

  1. Agglomerative, four clusters (0.2827)
  2. K-means, four clusters (0.3713)
  3. Agglomerative, three clusters (0.5303)
  4. Agglomerative, five clusters (0.5785)
  5. K-means, six clusters (0.5889)

- Other clustering methods have failed to achieve the minimum cluster size already with the number of clusters equal to 4 (K-medians, BIRCH) or even with the number of clusters equal to 3 (spectral).
- Spectral clustering appeared to be the worst approach based on the business requirements, which, combined with the best results of this method when analyzing the IDB index, calls into question the usefulness of the IDB index for analyzing clustering methods.

Analogous studies were also carried out for a doubled set of input data, and the results are presented in Table 5.5.

Table 5.5 Cluster sizes (261774 sessions and 48789 customers)

Clusters	Algorithm	Biggest	Smallest	Std. dev.	CSI
3	Agglom.	24150	9966	**7224.4404**	0.5873
	BIRCH	24836	10818	7514.2777	0.5644
	K-means	24836	11478	7441.1524	0.5378
	K-medians	24723	11897	7327.8371	**0.5188**
	Spectral	48785	2	28164.8782	0.9999
4	Agglom.	14673	9596	**2794.5534**	**0.3460**
	BIRCH	24836	6305	8663.4542	0.7461
	K-means	14911	9037	3133.2305	0.3939
	K-medians	27706	*1602*	11910.5431	0.9422
	Spectral	35204	2	16620.0036	0.9999
5	Agglom.	14554	5908	3117.6783	0.5941
	BIRCH	24836	2647	9011.3704	0.8934
	K-means	13081	7210	**2408.9150**	**0.4488**
	K-medians	29015	*904*	12330.1708	0.9688
	Spectral	34909	2	15223.6858	0.9999
6	Agglom.	10026	5908	2342.2112	0.5484
	BIRCH	24836	*2433*	8524.8264	0.9020
	K-means	10331	5334	**1851.5936**	**0.4837**
	K-medians	26647	*322*	10020.6601	0.9879
	Spectral	34550	2	14009.4032	0.9999
7	Agglom.	10026	4103	2550.4772	**0.5908**
	BIRCH	24836	*659*	8327.5251	0.9735
	K-means	9971	3454	**2370.1586**	0.6536
	K-medians	24935	*777*	8466.9825	0.9929
	Spectral	48116	*1*	18144.4274	0.9999
8	Agglom.	10026	*1537*	3148.1379	0.8467
	BIRCH	24836	*659*	7945.9267	0.9735
	K-means	9971	3442	**2437.1860**	**0.6548**
	K-medians	26473	*174*	8913.7341	0.9934
	Spectral	48036	*1*	16945.8898	0.9999

This time the results were not as clear-cut as before, but the main findings were similar. It turns out that K-means clustering gave the best results for the number of clusters from 5 to 8 (although for seven clusters the agglomerative approach gave a slightly better CSI value). With four clusters fixed, the best results were obtained for agglomerative clustering. With three clusters, the best indicator values were obtained for agglomerative clustering (SD) and K-medians (CSI).

The lowest CSI values were obtained for the following combinations [clustering method—number of clusters] in turn:

1. Agglomerative, four clusters (0.3460)
2. K-means, four clusters (0.3939)
3. K-means, five clusters (0.4488)
4. K-means, six clusters (0.4837)
5. K-medians, three clusters (0.5188)

It can be seen that the first two points are the same as in the first iteration of the study. It can also be seen that, once again, spectral clustering is not suitable for preparing groups of users for dedicated UIs, since the size of the clusters varies greatly. On the other hand, some hope can be pinned on the BIRCH method, which, although it did not give the best values for the indicators analyzed, up to five clusters gave business-acceptable results (the size of the smallest cluster being over 5% of the population).

5.2.3   Conclusion

The analysis of different combinations [clustering method—number of clusters] was carried out in two iterations, for different sets of input data. The results made it possible to answer the questions posed before the start of the study.

(**Q1.1**) *What criteria should determine the choice of a clustering algorithm for serving dedicated UIs?*

A set of criteria to be considered when selecting a clustering algorithm for serving a multivariant user interface in e-commerce has been reviewed. For a complete analysis of the various options, the following factors need to be taken into account:

- The duration of clustering (with fixed computing resources) is measured by the clustering time (without preprocessing) or by the cost of the resources required to process a given set of data (if it is possible to scale up computing resources)
- Clustering quality metrics—aggregated (CSI) and partial (Silhouette, Calinski–Harabasz) excluding IDB (due to inconsistency of results with business context indicators) but with the option to use additional metrics, e.g., Dunn Index
- Business context metrics, describing the size of the resulting clusters—COE and SD, which generally give convergent results

(Q1.2) *How to analyze different clustering methods before selecting an algorithm for multivariant UIs?*

The study's methodology, involving clustering for various combinations [clustering method—number of clusters], enables a detailed verification of the obtained results. However, this approach may prove time-consuming when considering numerous combinations. Therefore, initial reductions in the number of analyzed algorithms or refinement of resulting clusters may be necessary. Algorithm selection criteria may include performance considerations or previous clustering experience. Business analysis and decisions on the number of dedicated interface variants may also influence the limitation of resulting clusters. The study revealed that K-means and agglomerative clustering are key methods to consider, with agglomerative being more computationally intensive. Moreover, the adaptation parameters to the business context are expected to deteriorate with an increasing number of clusters.

(Q1.3) *What results obtained should exclude the clustering method from further analysis?*

Analyzing different [clustering method—number of clusters] combinations, starting with the smallest cluster number, allows for early rejection of unsuitable methods if clustering duration approaches or exceeds thresholds or if the smallest cluster size falls below accepted thresholds. Clustering methods failing to meet boundary conditions with fewer clusters are likely to fail with an increased cluster count. The inability to control the number of clusters beyond a preset threshold is a disqualifying feature, making DBSCAN less useful for serving dedicated UI variants.

(Q1.4) *Which clustering method should be selected for further research on the effectiveness of multivariate UI?*

The choice of clustering method should depend on the results of the analysis of the different factors, in particular the values of the different indicators. Given the business context, it is worth considering selecting the clustering method that gives the best results in terms of meeting the business requirements (e.g., CSI), provided that the method selected is acceptable in terms of computational complexity and available computing resources. It should be noted that at the time of the study, the recommendations from the clustering quality indicators used and the indicators assessing convergence with business requirements gave very similar results, in most cases indicating the same clustering methods and the same number of resulting clusters. This means that the possible prioritization of clustering quality indicators (e.g., COE) should not significantly change the recommendation of clustering parameters.

The research carried out also makes it possible to achieve the objectives set.

(G1.1) *To analyze implemented clustering algorithms for suitability in clustering customers for serving multivariant UIs*

All implemented clustering algorithms (agglomerative, BIRCH, DBSCAN, K-means, K-medians, Spectral) underwent analysis. Evaluation of the combination [clustering method—number of clusters] requires consideration of various dimensions that can determine the choice of clustering parameters for a specific implementation. One of the dimensions analyzed was alignment with the business context—clustering customers to serve a multivariant UI. To compare different options, a CSI indicator has been proposed as an alternative to typical clustering quality indicators.

(**G1.2**) *To verify the method for selecting the optimal clustering algorithm*
The methodology for selecting optimal clustering parameters was applied in an e-shop, confirming its validity. The research also identified factors optimizing the procedure for determining clustering parameters, such as reducing analyzed options or limiting analysis to selected indicators describing clustering results.

(**G1.3**) *To select the best (or best) clustering algorithm(s) for further research on the effectiveness of multivariate UIs*
The results of the study indicate that for the collected e-commerce user behavior data, in the context of verifying the effectiveness of dedicated user interfaces, the optimal clustering parameters are:

1. Agglomerative clustering with 4 clusters (Fig. 5.4)
2. K-means clustering with 4 clusters (Fig. 5.5)

Both recommended options consistently ranked among the best in each analysis and exhibited superior values for indicators describing compliance with the business context. No single approach was chosen to enable a comparison of the clustering method's impact on results when assessing the effectiveness of serving dedicated UI variants. Notably, agglomerative clustering is more computationally intensive, making the K-means method preferable for large datasets.

Further research related to the selection of the optimal approach to choosing a clustering method for serving multivariate UIs could focus on the analysis of other clustering methods, including those that can potentially highlight the business context of clustering and methods that improve the performance of the solution. In

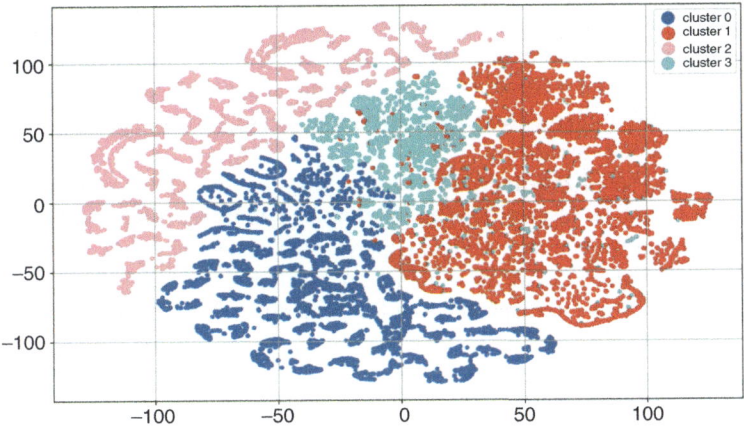

Fig. 5.4 TSNE cluster visualization—agglomerative, four clusters

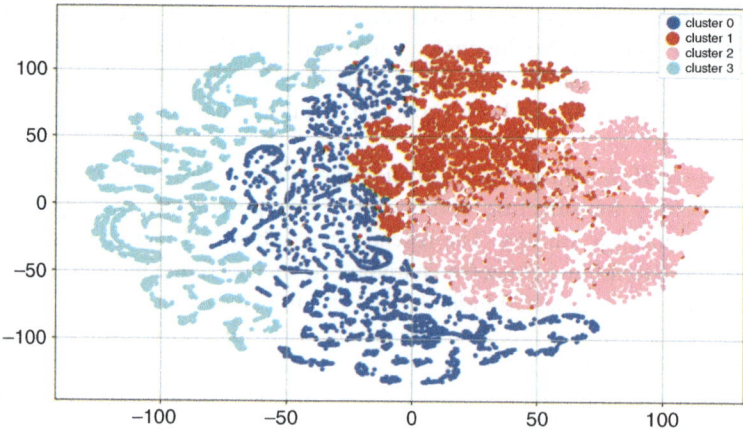

Fig. 5.5 TSNE cluster visualization—K-means, four clusters

addition, it would be worthwhile to verify the consistency of the results obtained with the results of tests on datasets with the same structure from different sources (e-shops). This would allow the generalization of the conclusions and a broader validation of the proposed method for selecting clustering parameters.

5.3 Analysis of the Effectiveness of the Interface Variants

5.3.1 Introduction

Serving a multivariant user interface is not widely used in practice. One of the reasons for this is uncertainty about the economic sense of investing in such a solution. On the one hand, personalization is a trend that is increasingly influencing the development of e-commerce, and AI applications are raising hopes of seeing dependencies that cannot be seen using traditional approaches to customer analysis; on the other hand, implementing a mechanism to serve a multivariant UI requires a measurable financial investment.

The use of personalized interfaces can lead to many benefits, such as the ability to better tailor the UI to user groups and increase customer satisfaction, but it should also deliver tangible benefits, such as increased CR or increased AOV, to make the implementation economically viable. Accordingly, several studies have been conducted to verify the impact of providing a dedicated interface on e-commerce KPIs.

Goals

The objectives of the research into the impact of providing dedicated user interfaces on e-commerce business efficiency were to:

(G2.1):   Verify the existence of a relationship between the interface variant served and the values of e-commerce performance indicators.

(G2.2):   Verify the impact of the chosen clustering method on the efficiency of the dedicated user interface.

(G2.3):   Identify the determinants of the economic rationality of serving multivariant interfaces.

Research Questions

To achieve the stated objectives of the study, the following research questions were addressed:

(Q2.1):   Can serving a dedicated interface affect e-commerce KPIs?

(Q2.2):   Will the same dedicated interface variant be as effective for different groups (clusters) of customers?

(Q2.3):   Is it possible to configure such a dedicated interface variant so that key performance indicators have higher values?

(Q2.4):   Can the choice of clustering method affect the effectiveness of a dedicated UI variant?

(Q2.5):   Can serving a dedicated interface variant to new customers affect the effectiveness of e-commerce?

Research Methodology

The research took the form of experiments carried out using the developed e-commerce multivariate user interface delivery platform.

To answer the research questions posed, a series of experiments were conducted. The scheme of the experiments was the same and included:

1. Collecting data about the e-shop's customer behavior
2. Customer clustering
3. Analysis of customer behavior in selected clusters
4. Selection of modifications to be implemented as part of a dedicated interface variant
5. Serving:

   a. Dedicated interface variant to the part of the customers in the cluster who have been assigned this interface variant
   b. Default interface variant to the part of the customers in the cluster who have been assigned this interface variant

6. Analysis of CR, AOV, and PCR values
7. Testing the hypotheses set and concluding

Different clustering methods were used to group the e-shop customers based on the learning dataset. The choice of algorithms and their parameters resulted from an analysis of the quality of the clustering described in the previous section and the objectives set for subsequent iterations of the research. Accordingly, the clustering method and the learning dataset were specified separately for each study when discussing the results obtained.

Clustering resulted in a set of customer identifiers that were assigned to specific clusters. The user behavior of each cluster for which a specific interface was to be provided was analyzed using the information on cluster characteristics, action reports, and action sequences. This formed the basis for selecting the modifications that made up the interface variant under investigation.

To conduct the study, selected customer clusters were divided into two groups. One group received the interface variant, while the other received the standard interface. In one study, proportions varied as two personalized user interface variants were prepared for a customer cluster. In this case, 1/3 of the users served as the test group for the first variant and 1/3 for the second variant, and 1/3 received the default user interface.

The data collected on customer behavior and decisions on consecutive days was used to calculate the values of selected performance indicators. Calculations were made independently for each group of customers surveyed. The values of the CR and AOV indicators were calculated using standard formulae, while the scoring of the following events was included in the calculation of the PCR metric:

- Switching from the homepage to the product listing—10 points
- Switching from the homepage to the product page—10 points
- Switching from the listing to the product page—10 points
- Switching from the product page to another product page—5 points
- Adding the product to the basket—20 points

Points were added according to the assumed expected customer journey in the e-shop.

The comparison of the results obtained within the cluster by users served by the dedicated UI variant and by users served by the standard interface was the basis for concluding the impact of the interface changes introduced on the business efficiency of the e-shop. A *two-proportion Z test* was used to verify the hypothesis on the conversion rates.

5.3.2 Study 1

Goals of the Study

(**G2.1.1**): To verify whether serving a dedicated interface has an impact on selected e-commerce performance indicators

(**G2.1.2**): To verify that agglomerative clustering with the four adopted clusters allows customers to be grouped in a way that allows the design of a dedicated layout that is better than the default user interface

To achieve the objectives, it was assumed that the verification would include the measurement of the CR indicator since modifications to the user interface were selected that were intended to improve the values of this indicator.

As CR is a proportion indicator, a research hypothesis was set:

$$H1_0 : cr_A = cr_s$$
$$H1_1 : cr_A > cr_s$$

where cr_A is a conversion ratio for the dedicated UI variant A (A) and cr_s is a conversion ratio for the standard (default) UI variant.

Basic Information on the Study

- The number of user sessions included in the clustering: **261774**
- Clustering method: **agglomerative**
- Number of clusters: **4**
- Cluster sizes (Table 5.6)
- Key cluster characteristics (Table 5.7)

For the study, **cluster 1** was chosen because it covered customers with the highest activity (action) and the highest return on visits (revenue). These factors were considered key due to:

- The higher probability that users from this cluster would revisit the online shop during the period of verifying the impact of the dedicated UI on customer behavior
- The potential benefits of increasing the value of e-shop performance indicators for the most active group of customers

For the selected cluster, an analysis of cluster characteristics and action reports and action sequences was performed, which allowed the selection of 13 modifications (Fig. 5.6) that were made to the default UI.

Table 5.6 Cluster sizes—experiment 1

	Cluster 0	Cluster 1	Cluster 2	Cluster 3
Size	14870	18080	9966	5873
Size proc.	30.4782%	37.0575%	20.4267%	12.0375%

Table 5.7 Key cluster characteristics—study 1

	Cluster 0	Cluster 1	Cluster 2	Cluster 3
Action	17.5074	93.7845	15.3367	36.7037
Event	3.1198	22.7626	4.3662	9.5326
Revenue	0.0844	10.2206	0.4798	2.5689
E-commerce order	0.0031	0.1312	0.0421	0.055
E-commerce abandoned cart	0.0052	0.5342	0.0001	0.2959
Search	0.1738	0.6267	0.1908	0.3465

Tested Areas	AREA IDENTIFIER	AREA VALUE	STATUS
	product_details_rating_star	RSTT	Accepted
	product_details_add_wishlist	AWLOL	Accepted
	cart_free_shipping	MITLC	Accepted
	search_input_header	HISS	Accepted
	footer_links	AEOD	Accepted
	menu_mobile_nav	MNBP	Accepted
	product_details_right_column	FIXED	Accepted
	product_details_header_sku	SPS	Accepted
	product_details_header_price	BPP	Accepted
	product_details_model_info	MIDG	Accepted
	product_details_size_info	LTSD	Accepted
	product_details_delivery_info	DIST	Accepted
	product_details_gallery	GDBFAT	Accepted

Fig. 5.6 Implemented modifications—experiment 1

Results of the Study

Verification of the impact of the dedicated interface variant was based on 65601 e-shop customer sessions, of which users from the selected cluster completed 5025 sessions (the dedicated interface was served during 2582 sessions and the default interface during 2443 sessions).

PCR indicator

- Dedicated UI interface: **45.77**
- Default UI interface: **46.67**

CR indicator

- Dedicated UI interface: **4.11%**—106 orders placed in the test period
- Default UI interface: **2.91%**—71 orders placed in the test period

AOV indicator

- Dedicated UI interface: **42.48**
- Default UI interface: **44.47**

CR*AOV indicator

- Dedicated UI interface: **1.74**
- Default UI interface: **1.29**

Verification of hypothesis H1

- Calculated Z-value: **2.3046**
- Critical Z-value for the level of significance $p = 0.05$: **1.64**
- Conclusion: **the hypothesis $H1_0$ should be rejected in favor of the alternative hypothesis $H1_1$**

Findings of the Study

The verification sample size was approximately 25% of the size of the learning dataset. During the experiment, it was possible to identify 177 orders placed by customers qualified in the cluster selected for the study, which can be considered a statistically significant sample. For the other clusters generated, there were far fewer orders placed (*cluster 0—31, cluster 2—14,* and *cluster 3—51*). Considering that the number of orders is crucial for the values of the CR and AOV indicators, it can be concluded that the time to verify the impact of interface options can be significantly increased for clusters containing inactive (and therefore presumably infrequent returning) customers.

The results obtained also make it possible to refer to the objectives of the experiment. In the case of the CR indicator (which is crucial in the analysis of the effectiveness of e-commerce), a significantly better result (by 41%) was observed for the designed dedicated interface variant (**G2.1.1**). The conclusion also confirms the rejection of hypothesis $H1_0$.

This suggests the possibility of implementing UI modifications that significantly alter the CR indicator. Furthermore, the chosen clustering method demonstrated the ability to generate clusters with an improved CR value when served with a dedicated interface (**G2.1.2**). The analysis carried out allowed the alternative hypothesis $H1_1$ to be accepted, according to which the CR for the dedicated interface was higher than the CR for the standard interface.

However, no significant differences were observed for PCR and AOV ratios, implying that the modifications made may not significantly impact these metrics. This observation was confirmed in subsequent studies. Since the design of the dedicated interface was aimed at optimizing the CR rate rather than the AOV and PCR rates, such a conclusion is not surprising.

The last important finding of the experiment was that a very large group of customers (more than 75%) who visited the e-store during the interface verification period were new customers who had not previously been clustered. This is important information, as the lack of knowledge about the previous behavior of these customers does not allow them to be served an interface variant dedicated to a specific cluster. However, the question arises as to whether the new (previously, unclustered) customers should not be treated as a separate supercluster that can also be served with a dedicated interface to improve performance indicators. This issue has been addressed in subsequent studies.

5.3.3   Study 2

Goals of the Study

(G2.2.1): To confirm the results obtained during the first experiment

(G2.2.2): To verify that a dedicated interface variant designed for a specific cluster is as effective for other customer clusters

(G2.2.3): To verify whether the interface variant dedicated to the selected cluster will significantly affect the performance indicators for new (unclustered) users

To meet the goals, the following research hypotheses were set:

$$H2_0 : cr_{Ac1} = cr_{s1}$$
$$H2_1 : cr_{Ac1} > cr_{s1}$$

where cr_{Ac1} is a conversion ratio for the dedicated UI variant A (A) in the first cluster analyzed and cr_{s1} is a conversion ratio for the standard UI variant in the first cluster.

$$H3_0 : cr_{Ac2} = cr_{s2}$$
$$H3_1 : cr_{Ac2} < cr_{s2}$$

where cr_{Ac2} is a conversion ratio for the dedicated UI variant A (A) in the second cluster analyzed and cr_{s2} is a conversion ratio for the standard UI variant in the second cluster.

$$H4_0 : cr_{An} = cr_{sn}$$
$$H4_1 : cr_{An} < cr_{sn}$$

where cr_{An} is a conversion ratio for the dedicated UI variant A (A) served to unclustered (new) users and cr_{sn} is a conversion ratio for the standard UI variant of unclustered users.

Basic Information on the Study

- The number of user sessions included in the clustering: **376901**
- Clustering method: **agglomerative**
- Number of clusters: **4**
- Cluster sizes (Table 5.8)
- Key cluster characteristics (Table 5.9)

This time, **cluster 3** (the most active group, similar to the previous experiment) and **cluster 1**, the next most active group of customers in the e-shop, were selected for the study.

Table 5.8 Cluster sizes—experiment 2

	Cluster 0	Cluster 1	Cluster 2	Cluster 3
Size	22190	23644	13909	10830
Size proc.	31.4426%	33.5029%	19.7087%	15.3458%

Table 5.9 Key cluster characteristics—study 2

	Cluster 0	Cluster 1	Cluster 2	Cluster 3
Action	17.5854	59.6326	14.0408	127.1202
Event	3.3153	16.5083	3.9895	34.2009
Revenue	0.0731	6.0581	0.4132	13.5235
E-commerce order	0.0027	0.1176	0.0073	0.2268
E-commerce abandoned cart	0.0055	0.371	0.0009	0.7146
Search	0.1753	0.373	0.2046	1.1543

It should be noted that the numbering of the clusters in the subsequent experiments is different (in the previous experiment the most active customers went to cluster number 1 and in the described experiment to cluster number 3). The size of the clusters has also changed (in the previous experiment the cluster of the most active customers contained more than 37% of all clustered users and in the current study more than 15% of the users) and their characteristics (previously the cluster of the most active customers had on average more than 93 actions and in the next study already more than 127 actions).

The dedicated interface variant remained unchanged from the previous experiment, but this time it was served to half of the clients from both selected clusters and half of the new clients who were not assigned to either cluster.

Results of the Study

Verification of the impact of the dedicated interface variant was based on 76384 e-shop customer sessions, of which:

- Users from cluster 3 completed 5542 sessions (the dedicated interface was served during 2797 sessions and the default interface during 2645 sessions).
- Users from cluster 1 completed 3828 sessions (the dedicated interface was served during 1911 sessions and the default interface during 1917 sessions).
- New users (unclustered) completed 45416 sessions (the dedicated interface was served during 22555 sessions and the default interface during 22861 sessions).

Verification of hypothesis H2

- Calculated Z-value (for cluster 3): **2.4318**
- Critical Z-value for the level of significance $p = 0.05$: **1.64**
- Conclusion: **the hypothesis $H2_0$ should be rejected in favor of the alternative hypothesis $H2_1$**

PCR indicator			
	Cluster 3	Cluster 1	New users
Dedicated	44.41	47.19	43.45
Default	39.48	55.66	44.00

Number of orders

	Cluster 3	Cluster 1	New users
Dedicated	113	56	714
Default	75	85	718

CR indicator

	Cluster 3	Cluster 1	New users
Dedicated	4.04%	2.93%	3.16%
Default	2.83%	4.43%	3.14%

AOV indicator

	Cluster 3	Cluster 1	New users
Dedicated	42.83	42.53	41.84
Default	42.63	45.51	41.28

CR*AOV indicator

	Cluster 3	Cluster 1	New users
Dedicated	1.73	1.21	1.18
Default	1.03	1.90	1.19

Verification of hypothesis H3

- Calculated Z-value (for cluster 1): **−2.4695**
- Critical Z-value for the level of significance $p = 0.05$: **−1.64**
- Conclusion: **the hypothesis $H3_0$ should be rejected in favor of the alternative hypothesis $H3_1$**

Verification of hypothesis H4

- Calculated Z-value: **0.1517**
- Critical Z-value for the level of significance $p = 0.05$: **1.64**
- Conclusion: **the hypothesis $H4_0$ cannot be rejected in favor of the alternative hypothesis $H4_1$**

Findings of the Study

In the experiment described, the learning dataset was 44% larger than before. This had a significant impact on the distribution of customers in the clusters and on the characteristics of the clusters themselves. However, despite the change in the parameters of the cluster with the most active customers, it was found that the results obtained were consistent with the results of the previous experiment. A CR value of 42.5% (previously, 41%) higher was observed in the group of customers served with the dedicated interface variant compared to the group served with the standard interface. For the PCR and AOV indicators, there were again no significant differences between the groups of customers in this cluster, even though they were served different UI variants. This confirmed the results of the first experiment (**G2.2.1**) and is consistent with the results of the H2 hypothesis verification.

Results for customers from a slightly less active cluster were intriguing. The same interface variant that significantly increased CR scores in the previous experiment's discussed cluster proved inferior to the standard interface in all aspects (PCR, CR, and AOV) this could be the lower number of orders placed by this group of customers, but it can still be assumed that the designed interface variant is not suitable for this cluster of users (**G2.2.2**). Furthermore, the testing of hypothesis H3, juxtaposed with the conclusions regarding hypothesis H2, shows that the same UI variant that positively influenced the conversion of users from cluster 3 (hypothesis H2 rejected) negatively influenced the conversion of users from cluster 1 (hypothesis H3 rejected but with the alternative hypothesis reversed). This suggests that dedicated UI variants should be designed for specific customer groups based on an analysis of their behavior specifics, necessitating varied interface variants for different e-commerce customer clusters in subsequent experiments.

In the case of the last group of customers studied—new, unclustered users—there were practically no differences between the values of the PCR, CR (evidenced by no reason to reject hypothesis H4), and AOV indicators for the interface designed for active customers and the standard interface. On the one hand, such results confirmed that the same interface variant can have different efficiencies for different groups of customers (**G2.2.2**), but on the other hand, they did not provide a clear answer to the question of whether it is possible to design an interface variant that would change the values of the performance indicators for those customers who were not previously clustered (**G2.2.3**). This issue was further explored in subsequent studies.

5.3.4   Study 3

Goals of the Study

(**G2.3.1**): To verify whether the different variants of dedicated user interfaces, in different clusters, have the same effect on the proportion of orders placed

(**G2.3.2**): To verify whether it is possible to design an interface variant that would significantly influence the behavior of new customers

(**G2.3.3**): To verify whether changing the clustering method to k-means will affect the effectiveness of the previously designed variant of the dedicated interface

To meet the goals, the following research hypotheses were set:

$$H5_0 : \frac{op_{Ac1}}{op_{sc1}} = \frac{op_{Bc2}}{op_{sc2}}$$
$$H5_1 : \frac{op_{Ac1}}{op_{sc1}} \neq \frac{op_{Bc2}}{op_{sc2}}$$

where op_{Ac1} is the number of orders placed when the dedicated UI variant A (A) was served and op_{sc1} is the number of orders placed when the standard UI was served in the first cluster analyzed, and op_{Bc2} is the number of orders placed when the dedicated UI variant B (B) was served and op_{sc2} is the number of orders placed when the standard UI was served in the second cluster analyzed.

$$H6_0 : cr_{Cn} = cr_s$$
$$H6_1 : cr_{Cn} \neq cr_s$$

where cr_{Cn} is a conversion ratio for the dedicated UI variant C (C) served to unclustered (new) users and cr_s is a conversion ratio for standard UI variant.

$$H7_0 : cr_{Ac1-agg} = cr_{Ac1-kmeans}$$
$$H7_1 : cr_{Ac1-agg} \neq cr_{Ac1-kmeans}$$

where $cr_{Ac1-agg}$ is a conversion ratio for the dedicated UI variant A (A) with clusterization based on agglomerative method and $cr_{Ac1-agg}$ is a conversion ratio for the dedicated UI variant A (A) with clusterization based on K-means method; in both cases, similar clusters of the most active customers are analyzed.

Basic Information on the Study

- The number of user sessions included in the clustering: **441342**
- Clustering method: **k-means**
- The number of clusters: **4**
- Cluster sizes (Table 5.10)
- Key cluster characteristics (Table 5.11)

In this study, customers in **cluster 0** (the most active group) were left with the previously designed interface variant that improved the CR rate. Other interface variants were designed for the remaining clusters of customers and new (unclustered) customers (Fig. 5.7). For the study's purposes, users within each cluster were divided in half, with one subgroup receiving a dedicated interface and the other a standard interface. The same rule applied to new customers, with 50% serving the dedicated interface and 50% the standard interface.

Table 5.10 Cluster sizes—study 3

	Cluster 0	Cluster 1	Cluster 2	Cluster 3
Size	19395	18998	27277	28088
Size proc.	20.6862%	20.2628%	29.0930%	29.9580%

Table 5.11 Key cluster characteristics—experiment 3

	Cluster 0	Cluster 1	Cluster 2	Cluster 3
Action	125.5319	14.1690	53.9518	17.5373
Event	37.5710	4.1705	16.7635	3.2924
Revenue	13.1089	0.3937	5.5892	0.0722
E-commerce order	0.2162	0.0070	0.1105	0.0025
E-commerce abandoned cart	0.6622	0.0007	0.3653	0.0049
Search	0.9558	0.2112	0.4659	0.1777

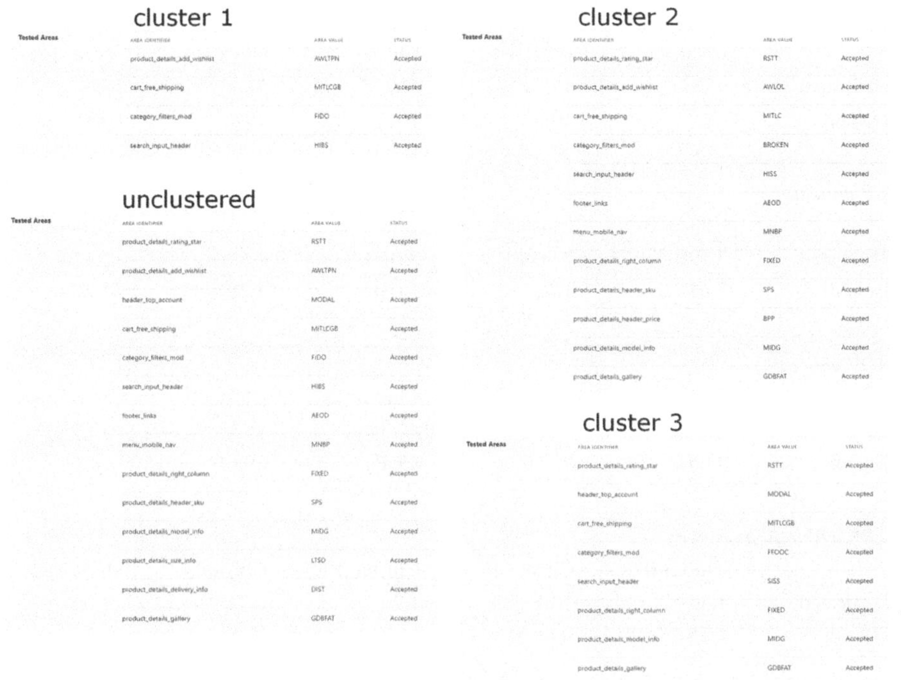

Fig. 5.7 Configuration of dedicated UI variants

Results of the Study

Verification of the impact of the dedicated interface variant was based on 89370 e-shop customer sessions, of which:

- Users from cluster 0 completed 5932 sessions (the dedicated interface was served during 2974 sessions and the default interface during 2958 sessions).
- Users from cluster 1 completed 1190 sessions (the dedicated interface was served during 587 sessions and the default interface during 603 sessions).
- Users from cluster 2 completed 6848 sessions (the dedicated interface was served during 3520 sessions and the default interface during 3328 sessions).
- Users from cluster 3 completed 3139 sessions (the dedicated interface was served during 1653 sessions and the default interface during 1486 sessions).
- New users (unclustered) completed 47012 sessions (the dedicated interface was served during 23439 sessions and the default interface during 23573 sessions).

PCR indicator					
	Cluster 0	Cluster 1	Cluster 2	Cluster 3	New users
Dedicated	46.67	31.18	35.81	30.18	38.43
Default	42.56	26.63	37.23	28.56	39.14

Number of orders

	Cluster 0	Cluster 1	Cluster 2	Cluster 3	New users
Dedicated	95	17	106	26	579
Default	78	10	101	19	650

CR indicator

	Cluster 0	Cluster 1	Cluster 2	Cluster 3	New users
Dedicated	3.19%	2.90%	3.01%	1.57%	2.47%
Default	2.64%	1.66%	3.03%	1.28%	2.76%

AOV indicator

	Cluster 0	Cluster 1	Cluster 2	Cluster 3	New users
Dedicated	41.82	34.80	42.34	45.11	40.57
Default	43.64	41.76	45.48	30.61	38.11

CR*AOV indicator

	Cluster 0	Cluster 1	Cluster 2	Cluster 3	New users
Dedicated	1.34	1.01	1.28	0.71	1.00
Default	1.15	0.69	1.38	0.39	1.05

Verification of hypothesis H5

- Calculated Z-value (for cluster 0 and cluster 2): **0.2483**
- Critical Z-value for the level of significance $p = 0.05$: **1.96**
- Conclusion: **hypothesis $H5_0$ cannot be rejected in favor of the alternative hypothesis $H5_1$**

Verification of hypothesis H6

- Calculated Z-value (unclustered users): **−1.9510**
- Critical Z-value for the level of significance $p = 0.05$: **1.96**
- Conclusion: **formally hypothesis $H6_0$ cannot be rejected in favor of the alternative hypothesis $H6_1$ but the calculated Z-value is very close to rejecting hypothesis $H6_0$**

Verification of hypothesis H7

- Calculated Z-value (cluster 3 from Study 2 for agglomerative clustering and cluster 0 for K-means clustering): **1.7725**
- Critical Z-value for the level of significance $p = 0.05$: **1.96**
- Conclusion: **the hypothesis $H7_0$ cannot be rejected in favor of the alternative hypothesis $H7_1$**

Findings of the Study

The study revealed that k-means clustering results were very similar to those of agglomerative clustering in terms of values characterizing resulting clusters. This was partially confirmed by the verification of hypothesis H7. In this experiment, each customer cluster had a dedicated interface variant, making it a target application of a multivariant e-commerce user interface. Interface variants differed in the number and nature of modifications, enabling a rough assessment of the designed dedicated interfaces. Results indicated that three interface variants improved CR values (cluster 0, cluster 1, and cluster 3—**G2.3.1**), one showed no significant differences in CR rate between dedicated and standard interfaces (cluster 2), and one demonstrated worsened conversion rates for unclustered users. Hypothesis H5 verification did not rule out the same effect of two different UI variants in two different customer clusters on the proportion between orders placed in the research and comparison groups, despite a more than 20% difference in absolute conversion rates for the comparison groups in the two clusters (with comparable conversion rates for both dedicated user interface variants).

Nevertheless, two additional observations are worth noting here:

• Higher PCR scores in the same clusters where CR scores were successfully increased
• Lower AOV values for most dedicated interfaces, but the CR*AOV values were still higher for the three interface variants

However, some of the results obtained should be treated with some caution. For user clusters with less activity, the number of orders placed was relatively low (27 for cluster 1 and 45 for cluster 3), so the results obtained cannot be considered statistically significant.

At first sight the results of the study showed that the designed interface variant for new customers was inferior to the default variant, meaning that one of the objectives (**G2.3.2**) was indeed achieved (the possibility of influencing the behavior of new users with a dedicated interface was observed), but the result obtained was the opposite of business expectations (the dedicated interface was worse than the default one). It was, therefore, necessary to find such a configuration of the dedicated UI that would positively influence the behavior of unclustered customers and to verify it in the next study. It should be noted that the formal hypothesis H6 of no effect of the dedicated UI variant could not be rejected because the calculated Z-value was 0.01 below the accepted threshold. However, from a practical point of view, a difference of more than 10 percent in conversion rate should be considered business-relevant.

Though overall cluster characteristics from agglomerative clustering and k-means clustering were similar, a slightly worse increase in CR values was observed for the most active customer group (41–42.5% with agglomerative clustering, 21% with k-means clustering) under the same interface variant throughout the study. Furthermore, the $H7_0$ hypothesis could not be rejected, suggesting that none of the clustering methods is superior. From a business perspective, it would be possible to tentatively conclude that agglomerative clustering gave a better assignment of customers to clusters in terms of their ability to increase the number of orders they place (**G2.3.3**), but the verification should be repeated in subsequent studies to increase its reliability.

5.3.5   Study 4

Goals of the Study

(G2.4.1): Verify the possibilities and effects of serving multiple dedicated interface variants to one group of customers

(G2.4.2): To reverify the feasibility of designing an interface variant that could improve the behavior of new users

To meet the goals, the following research hypotheses were set:

$H8_0 : cr_{Dn} = cr_s$
$H8_1 : cr_{Dn} > cr_s$

where cr_{Dn} is a conversion ratio for the dedicated UI variant D (D) served to unclustered (new) users and cr_s is a conversion ratio for standard UI variant.

$H9_0 : cr_{En} = cr_s$
$H9_1 : cr_{En} > cr_s$

where cr_{En} is the conversion ratio for the dedicated UI variant E (E) served to unclustered (new) users and cr_s is the conversion ratio for the standard UI variant.

Basic Information on the Study

• The number of user sessions included in the clustering: **523108**

In this study, clustering was solely employed to identify customers during the learning data collection. The results were not utilized to analyze the impact of interface variants on clustered customers.

To design a new dedicated interface variant, a UX analysis of the available modifications and their potential impact on unclustered customers was performed. Following expert analysis, ten modifications were selected as required changes, and four additional modifications were selected as options. The selection criterion for selecting the modifications was their potential impact on conversion (CR indicator), given the importance of this measure for the efficiency of the e-commerce business, if necessary, analogous efficiency. Analogous UX analyses can be carried out and conducted for other performance indicators (e.g., PCR, AOV), and interface modifications can be designed and verified in the same way as in similarly to the study described.

The analysis conducted resulted in the design of two variants of dedicated interfaces (Fig. 5.8)—a minimum version (10 modifications) and a maximum version (14 modifications). During the survey, 1/3 of the customers were served a minimum dedicated variant, 1/3 a maximum dedicated variant, and 1/3 a standard UI variant.

Results of the Study
The verification of the impact of the dedicated interface variant was based on 55622 e-shop customer sessions. The lower number of sessions included in the study was

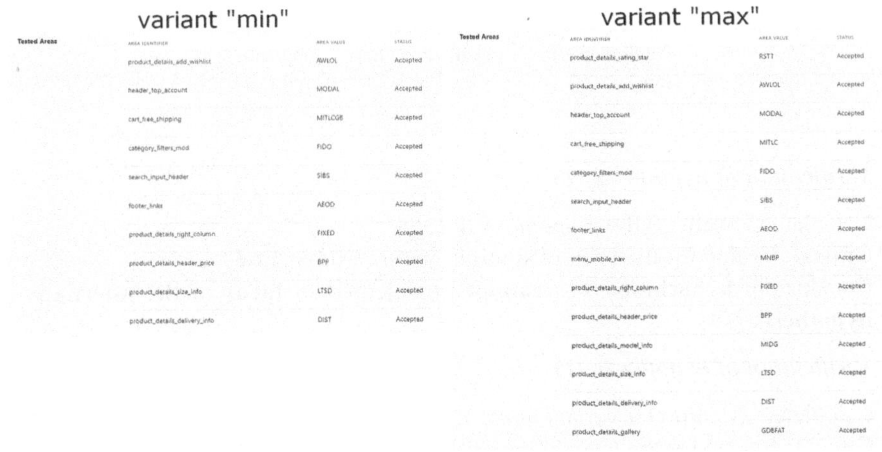

Fig. 5.8 Configuration of dedicated UI variants for unclustered users

due to the large number of unclustered customers, which allowed a sufficient number of orders to be placed relatively quickly. During the study:

- The dedicated interface in the minimum version was served during 8523 sessions.
- The dedicated interface in the maximum version was served during 8571 sessions.
- The default interface was served during 8852 sessions.

PCR indicator			
	Dedicated "min"	Dedicated "max"	Default
Mean	37.52	36.94	36.94
Std. dev.	66.61	68.57	64.57

Number of orders		
Dedicated "min"	Dedicated "max"	Default
208	210	192

CR indicator		
Dedicated "min"	Dedicated "max"	Default
2.44%	2.45%	2.17%

AOV indicator		
Dedicated "min"	Dedicated "max"	Default
32.67	45.76	36.10

CR*AOV		
Dedicated "min"	Dedicated "max"	Default
0.80	1.12	0.78

Verification of hypothesis H8

- Calculated Z-value (UI variant *max*): **1.2356**
- Critical Z-value for the level of significance $p = 0.05$: **1.64**
- Conclusion: **hypothesis $H8_0$ cannot be rejected in favor of the alternative hypothesis** $H8_1$

Verification of hypothesis H9

- Calculated Z-value (UI variant *min*): **1.1927**
- Critical Z-value for the level of significance $p = 0.05$: **1.64**
- Conclusion: **the hypothesis $H9_0$ cannot be rejected in favor of the alternative hypothesis** $H9_1$

Findings of the Study

The study described here was different from previous ones. This time, two versions of the dedicated interface were designed and delivered (along with the standard interface) to a group of customers. Furthermore, this was a group of customers whose behavior was unknown, as they had not been clustered before. In a sense, the verification carried out was in the nature of an A/B test (or in this case even an A/B/C test, as two versions of the dedicated UI were served—**G2.4.1**), the purpose of which was to identify the best (from the point of view of selected indicators) interface variant for the group of new users. This approach represents another business case for using the platform to serve a multivariant e-commerce user interface, without directly using the collected customer behavior information and its analysis to group customers.

The results of the study show that the interface variant has virtually no effect on the values of the PCR indicator. This means that the customer's journey in the e-shop, within the range covered by the indicator, does not differ despite the modification of the interface. It can therefore be assumed that the changes made do not affect the way new customers use the e-shop. The reason for this may be that the UX analysis and the implemented changes focus on increasing conversions (CR rate) rather than on changing the use of the e-shop.

From the standpoint of the study's objective, the most crucial results stem from the CR index analysis. It was observed that CR index values for both dedicated interface variants were approximately 13% higher than for the standard interface (**G2.4.2**). This is a result that seems satisfactory from the point of view of e-commerce business efficiency, as it allows the number of orders to increase significantly. This means that the objective has been achieved and the possibility of designing interface variants that improve the behavior of new users in terms of their purchasing decisions has been confirmed. However, it should be emphasized that, as there are no reasons to reject hypotheses $H8_0$ and $H9_0$, it cannot be statistically excluded that conversion rates for dedicated interface variants are not significantly higher than for the standard user interface.

Considering the business efficiency of the tested user interface variants, attention should be directed to the AOV indicator. In this case, the indicator varied significantly for the *min* and *max* variants. Orders placed in the *max* variant of the dedicated interface had a value more than 26% higher than orders placed when serving the standard interface, while orders in the *min* variant had a value of 10% lower. These results directly impact the potential benefits of implementing the studied modifications (CR*AOV ratio). It shows that the *max* variant of a dedicated interface generates more than 40% more average revenue per customer visit to the e-shop than the other considered interface variants. This not only confirms the achievement of the study's objective (**G2.4.2**) but also suggests that analyzing the CR indicator alone may not suffice to verify the benefits of implementing a multivariant interface.

5.3.6   Conclusion

The four-part study facilitated the verification of the effectiveness of dedicated user interfaces and provided answers to the posed inquiries.

(**Q2.1**) *Can serving a dedicated interface affect e-commerce KPIs?*

The results demonstrate that e-commerce performance indicators, such as CR or AOV, hinge on the customer-facing user interface. Personalizing the UI accordingly could positively impact business outcomes within the e-shop. Additionally, the findings indicate that UI alterations have the potential to influence the values of the PCR indicator, as defined within the study. This measure does not restrict itself only to customers who have placed an order. Rather, it enables the examination of the entire journey of the customers during their visit to the store. E-commerce may hold a specific interest in tracking the customer journey, which includes the benefit of merely browsing a page or perusing content (e.g., an offer), independent of placing a formal order. In this scenario, PCR enables the evaluation of interface alterations and their impact on the optimization of the UI variant viewed by a particular customer group.

(**Q2.2**) *Will the same dedicated interface variant be as effective for different groups (clusters) of customers?*

The conducted research suggests that employing the same dedicated interface variant for different customer groups results in varying levels of effectiveness. This finding implies that the conventional e-commerce practice of implementing a single UI for all customers is suboptimal. Instead, UI modifications should be customized based on the specific behavior of each customer group, supporting the feasibility of a multivariant user interface concept in e-commerce as a progressive enhancement.

(**Q2.3**) *Is it possible to configure such a dedicated interface variant so that key performance indicators have higher values?*

Changes in customer behavior and decisions resulting from dedicated interfaces can have either positive or negative impacts. While negative changes are undesirable, they provide insights into what should be avoided. On the other hand, positive changes directly benefit businesses. The results indicate that a specific version of the user interface, developed through the analysis of cluster characteristics and user behavior, significantly enhances e-commerce performance indicators. Notably, the CR indicator was positively impacted, with customers served by the custom interface placing considerably more orders than those from the same cluster served by the standard interface. This phenomenon was consistent across different customer segments in each round of research.

(**Q2.4**) *Can the choice of clustering method affect the effectiveness of a dedicated UI variant?*

The selection of a clustering method holds significant importance for subsequent dedicated interface servicing. Based on previous studies, two clustering methods, agglomerative clustering and K-means, were evaluated for usability in terms of business context and clustering quality. The research concluded that both clustering methods were suitable, with a dedicated user interface positively impacting e-commerce efficiency. It is worth noting that while agglomerative clustering offered slightly improved results, the K-means method may be more practical due to its lower computational complexity.

(**Q2.5**) *Can serving a dedicated interface variant to new customers affect the effectiveness of e-commerce?*

New customers, those whose behavior has not been captured due to a lack of previous interaction with the e-shop, constitute an essential group that can benefit from a custom user interface. This group is substantial and crucial to target for loyalty and repeat business. A dedicated user interface for new customers could include functionalities facilitating their familiarization with the store, its layout and offer, along with supplementary data to enhance brand recognition or confidence in the retailer. The investigation demonstrated that such a UI must be appropriately developed for this customer group. Serving new customers with a dedicated user interface that had a positive impact on active customers did not yield the intended effect. Enhancing the effectiveness of the online shop requires developing modifications tailored to the specific needs of new customers. When implementing multivariate user interfaces, considering the group of new customers as a key component and treating them as a supercluster is essential for optimal results, differing from conventional clustering techniques.

The four iterations of research to verify the impact of dedicated user interfaces on e-commerce performance indicators successfully achieved the set objectives.

(**G2.1**) *To verify the existence of a relationship between the interface variant served and the values of e-commerce performance indicators.*

The relationship between interface variants and e-commerce indicators has been confirmed across several dimensions. Findings from the study suggest that a specialized user interface can potentially impact various e-commerce performance metrics positively or negatively. Additionally, optimizing one indicator may lead to a decrease in the value of another indicator. It is important to note that the process of selecting modifications for an interface variant is iterative, and the final version may result from trial and error. Despite thorough analysis of cluster characteristics and customer behavior, chosen modifications may prove ineffective or inadequate, necessitating reassessment and repetition of the study to ensure alterations are proceeding in the intended direction.

(**G2.2**) *To verify the impact of the chosen clustering method on the efficiency of the dedicated user interface.*

Based on a comparative study of various clustering methods used for dedicated UI interfaces, this research has verified the impact of two clustering techniques (K-means and agglomerative) on e-commerce performance metrics. At the evaluation stage of the clustering methods, no significant differences were found between the two chosen approaches. Further research allowed verification of their impact on e-commerce customer behavior and decisions. Both methods have been successful in creating clusters of customers that can be used to serve UI modifications to influence their experience. However, the challenge of computational complexity should not be disregarded, particularly with substantial quantities of learning data. Choosing the appropriate clustering methods based on quality indicators and alignment with the business context enhances the probability of affirmative outcomes when serving customized user interfaces. If the examinations of e-commerce efficiency had been conducted for all of the clustering algorithms initially contemplated, the outcomes achieved would likely have been less optimistic.

(**G2.3**) *To indicate the determinants of the economic rationality of serving multi-variant interfaces.*

Economic evaluation of the suggested plan for serving multivariate UI is vital for deciding on its implementation. It presents potential measurable gains by improving e-commerce performance indices (e.g., CR and AOV), coupled with less quantifiable, more intangible assets, like heightened customer satisfaction, repeat purchases, or increased product/brand recognition. On the flip side, there are expenses linked to the setup and upkeep of the system. When utilizing machine learning approaches, it is crucial to consider that the amount of data on customer behavior will expand over time, leading to an increase in the intricacy of calculations required to generate clusters. While economic evaluation should be a significant factor in decision-making, it should not be the sole determining factor. It is worth considering that personalization is a significant trend in the development of e-commerce systems and implementing such solution marketing is relevant as an advantageous element in the highly competitive e-commerce market.

5.4 Discussion

E-commerce, an indispensable part of the global economy, has undergone significant changes in recent decades, altering shopping habits and the way people
interact with online retail platforms. Personalization has played a pivotal role in
driving this transformation, with tailoring online shopping experiences to individual
preferences and needs emerging as a core strategy for e-commerce firms. Therefore,
it is imperative to address the question of how to differentiate, attract, and retain a
loyal customer base in a highly competitive sector such as e-commerce.

Tailoring the shopping experience leads to enhanced customer satisfaction and
influences customer decisions, thereby increasing sales and loyalty. The conventional *one-size-fits-all* approach is becoming inadequate as e-commerce users
demand personalized interactions that consider differences in preferences, behavior,
or needs.

The prevalent use of personalization in e-commerce is mainly attributable to
intensifying competition, augmented data accessibility and processing capabilities,
and shifts in consumer demands.

E-commerce is experiencing substantial growth in the number of retailers, posing
a challenge for companies to distinguish themselves from competitors. Personalization facilitates brands to differentiate themselves by generating memorable and
exclusive customer experiences. It has a functional dimension, making the e-shop
easier to use, and a marketing dimension, emphasizing the modernity of the message
to customers.

Modern ICT solutions enable the collection and analysis of substantial customer
data to comprehend individual preferences, behaviors, and purchasing practices.
This not only supports the provision of customized content and designs to customers
but also entails inherent risks. The critical issue is that of privacy and the limitations
posed by personal data collection. Customers are becoming more conscious of the
methods and locations for obtaining data on their online activities, no longer readily
accepting the uncontrolled tracking of their behaviors and decisions. Regarding
potential personalization, the failure to secure customer agreement to information
gathering can impede and even prevent personalization as recommendations may be
misguided, resulting in unintended consequences.

Contemporary consumers increasingly seek to differentiate themselves from the
masses while still desiring tailored treatment. As a result, seamless and personalized
online shopping experiences are now expected by many users. Brands that provide
bespoke product recommendations, content, and offers are more likely to attract
engagement, as individual preferences and needs are seen as customer-centric,
incentivizing purchases.

Personalization in e-commerce takes various forms, including personalized
advertising campaigns, product and price offers, loyalty programs, search results,
and tailored content. These solutions mainly personalize the content and not the
appearance of the message. Nonetheless, the presentation of the delivered content
can and should also be personalized. Meeting this challenge is possible by providing

customers of the e-shop with a multivariant user interface. Such a solution should dynamically tailor the online shopping experience for customer groups without compromising their privacy during data collection. As personalization of not only content but also design is not widely used in e-commerce, it was crucial to define the architecture for such a solution and verify its effectiveness in shaping customer behavior and decision-making.

In essence, a platform designed to offer multivariant e-commerce user interfaces is a specific application of recommender systems. As such, its operating principles are comparable to those of similar solutions. The primary components of this platform are data collection, data processing, recommendation generation, and impact evaluation. However, a closer look reveals that the proposed solution differs from the usual recommendation systems used in e-commerce.

The primary distinction can be observed in the scope of compiled data. Standard product recommendation systems focus on studying orders and products in the shopping basket, limiting the analyzed dataset. These systems often combine data from various shops to suggest recommendations like *other customers also bought*. Customer attributes, such as age, gender, and location, are also scrutinized. However, due to the trend toward anonymization of online activity, the future of this approach is uncertain. Over time, customers may refuse to provide such information or knowingly provide inaccurate data, reducing the effectiveness of customer knowledge-based solutions.

For a system designed to serve a multivariate user interface, relying solely on purchase decision information is insufficient. It is imperative to consider the entire customer journey within the e-store as a single entity, with all stages interconnected and dependent on both the user's visual experience and actions. The proposed solution captures information on every activity performed by the e-commerce customer, even those not resulting in product selection or purchase. This allows for a deep understanding of customer behavior, positively impacting subsequent recommendations. However, it involves collecting large amounts of data, requiring processing with appropriate algorithms. Compliance with regulations and trends related to customer information collection requires anonymization of identified behaviors and decisions.

The second key distinction concerns how the processed data affects the outcome. In e-commerce, recommendation systems typically generate lists of matching products, advertisements, or texts using a variety of approaches, most commonly collaborative filtering. In contrast, the proposed solution focuses on identifying links between users based on their behavior and creating clusters of individuals with similar characteristics. Analysis of the characteristics of each group leads to the selection of modifications that create a customized interface variant. The set of modifications can be determined by the expertise of a UX specialist or by activating the auto-adaptation mechanism, which verifies the impact of subsequent modifications and accepts those with the desired effect. Given this approach to data collection and analysis, where groups of customers are recommended, it must be assumed that the solution will only work within a single online store. The results are not directly transferable to other systems, which by their nature have a different user interface and therefore different customer behavior. This means that sufficient input data must be collected before the dedicated UI serving mechanism is deployed to reliably cluster customers.

Another difference lies in the use of the processed data results. With product recommendation systems, the results can be implemented directly by presenting a designated product list. However, accommodating a multivariant user interface requires additional effort. First, the behavior of customers in clusters has to be analyzed, and then the specified interface adjustments have to be designed and implemented. Therefore, it is advisable to anticipate possible modifications during the planning and execution stages of the complete e-commerce user interface. In the typical approach to e-commerce design, several proposals often have to be rejected to develop a single compromise solution. A multivariant user interface overcomes this problem by facilitating the implementation of different proposals and delivering the relevant adaptations to the customers who require them.

On the other hand, when monitoring the results of applied recommendations, a multivariant user interface approach can be evaluated similarly to traditional recommendation systems. The ultimate benchmark depends on measurable indicators associated with business outcomes, especially conversion rates and order values. For multivariant UIs, it is worth conducting a thorough efficiency analysis that includes indirect customer activity. A change to the interface can affect not only the final conversion but also the paths users take as they navigate through the e-shop.

The concept of multivariant e-commerce user interfaces complements other approaches to message personalization used in practice. However, unlike its counterparts, it has not been extensively researched or validated. Therefore, the designed mechanism was implemented and verified, and its impact on e-commerce performance indicators of dedicated interfaces was investigated.

The first stage of the investigation was to select a clustering algorithm to use as the main tool for analyzing the customer behavior data collected. There are various ways to arrange user clusters, but in the context of a particular business environment aiming to provide a consistent user interface to a specific group of e-commerce customers, not all methods are equally advantageous. It is important to consider both the distribution of users across the resulting clusters and the time and resources required generating the clusters. The complexity of designing and deploying a bespoke UI variant influences the desired number of clusters and minimum customer group sizes. Therefore, it is not recommended to have too many clusters or a limited number of customers in them, as the effort invested in developing the UI for these groups may not be compensated. The resources and time required for clustering are important considerations due to the significant amount of customer behavior data that needs to be processed. Dedicated service interfaces have an advantage over order-based analytics solutions in that they gain knowledge of every customer activity in the e-shop, requiring the collection of larger datasets. However, this entails potentially higher maintenance costs for the recommendation system.

Taking into account the business context of the research, the study identified the most effective clustering methods for grouping customers to serve dedicated UIs in practice. The results show that both agglomerative clustering and K-means clustering produced the best results; however, the former was more complicated and required more computational resources. Therefore, it can be concluded that the K-means algorithm is the optimal choice due to its greater scalability. As the analysis was only carried out on data from one e-commerce site, the results obtained are purely illustrative and cannot be directly extrapolated to other implementations. However, it is worth noting that the proposed method for analyzing and selecting the most appropriate clustering approach can be applied to other multivariate UI mechanisms. Possible verification of the results collected from additional datasets can serve as a basis for further investigation and the potential for drawing general conclusions.

The clustering parameter recommendation highlighted two methods that best met the accepted decision criteria. Both were used in the second part of the research, which aimed to determine the effect of specialized interface variations on e-commerce performance indicators. The results of the study confirmed that dedicated user interface modifications can positively influence customer behavior and decision-making, leading to tangible business benefits. The extent of these benefits may vary depending on the technical ability to implement the changes and the choice of modifications determined by the study of customer behavior within different clusters. The findings imply that to progress with the proposed mechanism, it may be necessary to increase the number of adaptable changes (thus increasing the variety of designated UI versions) and to use auto-adaptation. This would replace the need for expert knowledge in selecting an appropriate set of modifications.

An additional issue identified during the study relates to first-time visitors to the e-commerce platform whose behavior has not been analyzed. This represents a significant subgroup of e-commerce customers that must not be ignored when providing UI variants. Adapting to their specific needs, given their lack of familiarity with e-commerce navigation, products, or sellers, is a way to cultivate their loyalty. However, it is worth remembering to use modifications tailored to the needs of this group, as only such an approach can have the desired effect. The results of the study show that significant gains in performance metrics can be achieved by providing a unique user interface specifically for this customer group, rather than using the one designed for returning customers.

Multivariant user interfaces in e-commerce can be a valuable addition to the personalization methods commonly used. However, their implementation requires a different approach than solutions such as advertising or product recommendations. This is due to the need to collect more data and to design and prepare the interface modifications that make up the delivered variants. The upfront costs of implementing this approach may be higher than other personalization methods. However, the research suggests that this solution can bring tangible benefits to e-shop owners. In addition, the marketing potential and the opportunity to differentiate from competitors should not be underestimated—valid reasons to justify moving from a universal user interface to a customized one.

Reference

1. Grzech A, Juszczyszyn K, Kolaczek G, Kwiatkowski J, Sobecki J, Swiatek P, Wasilewski A (2014) Specifications and deployment of SOA business applications within a configurable framework provided as a service. Adv SOA Tools Appl. https://doi.org/10.1007/978-3-642-38957-3_2

Appendix A
Pseudocode

A.1 K-means

K-means Pseudocode [45]
```
KMeans(D, k, max_iterations)
    centroids = randomly_select_k_data_points(D)
    for iteration = 1 to max_iterations:
        clusters = assign_points_to_clusters(D, centroids)
        new_centroids = calculate_new_centroids(clusters)
        if centroids_converged(centroids, new_centroids):
            break
        centroids = new_centroids
    return clusters, centroids
```

where:

D: The dataset to be clustered

k: The number of clusters to create

$max\_iterations$: The maximum number of iterations to run the algorithm

$randomly\_select\_k\_data\_points(D)$ randomly selects k data points from the dataset as the initial centroids for the clusters.

$assign\_points\_to\_clusters(D, centroids)$ assigns each data point to the nearest centroid by computing the distance between each data point and each centroid; the data points are grouped into k clusters based on the nearest centroid.

$calculate\_new\_centroids(clusters)$ calculates the new centroids for each cluster based on the mean (average) of the data points in that cluster.

$centroids\_converged(centroids, new\_centroids)$ checks if the centroids have changed significantly between iterations—if not the algorithm stops.

© The Editor(s) (if applicable) and The Author(s), under exclusive license   127
to Springer Nature Switzerland AG 2024
A. Wasilewski, *Multi-variant User Interfaces in E-commerce*, Progress in IS,
https://doi.org/10.1007/978-3-031-67758-8

A.2   BIRCH

```
BIRCH Pseudocode [2]
BIRCH(T, B, N, S)
   CF = empty CF-tree
   for each data point X in the dataset:
       Insert(X, CF)
   Clustering(CF, N, S)
   OutputClusters(CF)

   Insert(X, CF)
   LeafNode = FindLeafNode(X, CF)
   if LeafNode has room for another point:
       InsertIntoLeaf(LeafNode, X)
   else if LeafNode can be split:
       Split(LeafNode, CF)
       Insert(X, CF)
   else:
       CreateNewLeafNode(LeafNode, X)

   FindLeafNode(X, Node)
   node = CF.root
   while node is not leaf:
       min_distance = infinity
       for each entry in node.entries:
           distance_to_entry = distance(entry.CF, X)
           if distance_to_entry < min_distance:
               min_distance = distance_to_entry
               nearest_entry = entry
       node = nearest_entry.pointer
    return node

   InsertIntoLeaf(LeafNode, X)
       LeafNode.entries.append(X)

   Split(LeafNode, CF)
   S1, S2 = CreateNewLeafNodes()
   for entry in LeafNode.entries:
       if distance(entry.CF, S1.CF) < distance(entry.CF, S2.CF):
           InsertIntoLeaf(S1, entry)
       else:
           InsertIntoLeaf(S2, entry)
   parent = LeafNode.parent
   if parent is not null:
       parent.entries.remove(LeafNode)
       parent.entries.append(S1)
       parent.entries.append(S2)
   return S1, S2
```

(continued)

```
CreateNewLeafNode(LeafNode, X)
new_leaf_node = new CF-leaf-node
new_leaf_node.entries.append(X)
parent = LeafNode.parent
if parent is not null:
    parent.entries.remove(LeafNode)
    parent.entries.append(new_leaf_node)
new_leaf_node.parent = parent
if CF.root == LeafNode:
    CF.root = new_leaf_node
LeafNode.parent = new_leaf_node

Clustering(CF, N, S)
L = GlobalClustering(CF, N, S)
for each entry in L:
    SubClustering(entry)

GlobalClustering(CF, N, S)
L = []
TraverseTree(CF.root, L)
SortEntries(L)
return L[N]

TraverseTree(node, L)
if node is leaf:
    for each entry in node.entries:
        L.append(entry)
else:
    for each entry in node.entries:
        TraverseTree(entry.pointer, L)

SortEntries(L)
// sorting logic based on CF-size, distance, etc

SubClustering(entry)
cluster = []
for each data point X in entry:
    if X is not clustered:
        C = NewCluster(X)
        MarkAsClustered(X)
        Neighbors = RegionQuery(X, S)
        if length(Neighbors) >= N:
            for each data point Y in Neighbors:
                if Y is not clustered:
                    MarkAsClustered(Y)
                    Neighbors_Y = RegionQuery(Y, S)
                    if length(Neighbors_Y) >= N:
                        Neighbors = Neighbors union Neighbors_Y
                    if Y not in cluster:
                        C.append(Y)
```

(continued)

```
            cluster.append(C)

NewCluster(X)
C = new cluster
C.centroid = X
return C

MarkAsClustered(X)
// mark X as clustered in the data structure

RegionQuery(X, S)
Neighbors = []
for each data point Y in CF
    if distance(X, Y) <= S:
        Neighbors.append(Y)
return Neighbors

OutputClusters(CF)
for each entry in CF:
// print the data points in the cluster
```

where:

T: Threshold for branching factor

B: Threshold for the number of points in a node

N: Threshold for the number of clusters to keep

S: Threshold for the maximum number of points in a subcluster

A.3 DBSCAN

```
DBSCAN Pseudocode [3]
DBSCAN(D, epsilon, min_points)
   for each data point P in D:
       if P is not visited:
           mark P as visited
           Neighbors = get_neighbors(P, epsilon, D)
           if len(Neighbors) >= min_points:
               create a new cluster C
               expand_cluster(P, Neighbors, C, epsilon, min_points)

expand_cluster(P, Neighbors, Cluster, epsilon, min_points):
add P to Cluster
for each data point P' in Neighbors:
    if P' is not visited:
```

(continued)

```
            mark P' as visited
            Neighbors' = get_neighbors(P', epsilon, D)
            if len(Neighbors') >= min_points:
                Neighbors = Neighbors joined with Neighbors'
        if P' is not yet a member of any cluster:
            add P' to Cluster

    get_neighbors(P, epsilon, D):
    return all data points in D within epsilon distance of P
```

where:

D (dataset) is the input data.

$epsilon$ is the neighborhood radius.

min_points is the minimum number of points required to form a cluster.

References

1. Zhang T, Ramakrishnan, Livny M (1997) BIRCH: a new data clustering algorithm and its applications. Data Min Knowl Discov. https://doi.org/10.1023/A:1009783824328
2. Guo D, Chen J, Chen Y, Li Z (2018) LBIRCH: an improved BIRCH algorithm based on link. In: ICMLC 2018: proceedings of the 2018 10th international conference on machine learning and computing. https://doi.org/10.1145/3195106.3195158
3. Kotary DK, Nanda SJ (2021) A distributed neighbourhood DBSCAN algorithm for effective data clustering in wireless sensor networks. Wirel Pers Commun. https://doi.org/10.1007/s11277-021-08836-y

Appendix B
User Interface Modifications

No	Identificator	Area	Description
1	RSTT	product_details_rating_star	Rating information moved above the product name—change for both desktop and mobile; stars are enlarged
2	AWLOL	product_details_add_wishlist	Add Wishlist as inline icon with Add to Cart—mobile view common (except for enlarged icon)
3	AWLTPN	product_details_add_wishlist	Add Wishlist in one line with the product name—common mobile view
4	MODAL	header_top_account	Popup displays when link is clicked; popup displays login form and registration link
5	MITLC	cart_free_shipping	Free delivery information in the left column—on a white background; when the free delivery threshold is reached—the background of the box with the information is green
6	MITLCGB	cart_free_shipping	Free delivery information in the left column—on a gray background; when the free delivery threshold is reached—the background of the information box is green
7	FIDO	category_filters_mod	Changing the order of the filters: category (if any)—price—size—colour—other filters
8	FFOOC	category_filters_mod	Only the first filter expanded by default—other filters collapsed by default

(continued)

No	Identificator	Area	Description
9	BROKEN	category_filters_mod	Broken filter view: categories—in the left column, other filters—moved above the content with products; modifications: filter with categories remains in the left column in classic form, filters in the header in dropdown form (when selected, the list will collapse an additional window), and active filters have been repainted
10	HISS	search_input	Hidden input field for writing, field only visible after clicking on magnifying glass
11	SISS	search_input	Input field shown
12	HIBS	search_input	Hidden input field, magnifying glass icon enlarged (25x25 px)
13	SIBS	search_input	Input field shown, magnifying glass icon enlarged (25x25 px)
14	AEOD	footer_links	Footer still developed on desktop, for mobile additional arrows in accordion
15	MNBP	menu_mobile_nav	Mobile menu—navigation at the bottom of the page
16	FIXED	product_details_right_column	Right column is fixed when scrolling down the page
17	SPS	product_details_header_sku	Product SKU above the product name
18	BPP	product_details_header_price	Enlarged product price
19	MIDG	product_details_model_info	Model information moved to the left column, underneath the photo galleries—independent of the gallery design
20	LTSD	product_details_size_info	Link to size chart moved below sizes
21	DIST	product_details_delivery_info	Shortened delivery text, deleted phrase *Produkts pieejams (Product available)*
22	TMTL	product_details_tabs	The tabs on the product card have been moved to the left-hand side—under the galleries, regardless of the gallery design
23	TCCR	product_details_tab	The tabs on the product card rolled up in the right-hand column
24	TCCL	product_details_tab	The tabs on the product card moved to the left side—under the galleries, regardless of the gallery design
25	GDBFAT	product_details_galery	As part of the changes for the desktop: removal of the second large photo; as part of the changes for mobile devices: main photo enlarged to the width of the screen and added thumbnails under the photo (instead of bullet)
26	RED	category_widget_area	Buttons in red—in category listings above products

SPRINGER NATURE

GPSR Compliance

The European Union's (EU) General Product Safety Regulation (GPSR) is a set of rules that requires consumer products to be safe and our obligations to ensure this.

If you have any concerns about our products, you can contact us on ProductSafety@springernature.com

In case Publisher is established outside the EU, the EU authorized representative is:

Springer Nature Customer Service Center GmbH
Europaplatz 3
69115 Heidelberg, Germany

The manufacturer's authorised representative in the EU is Springer
Nature Customer Service Centre GmbH, Europaplatz 3, 69115 Heidelberg,
Germany. If you have any concerns regarding our products, please
contact ProductSafety@springernature.com

Printed and bound by CPI Group (UK) Ltd, Croydon, CR0 4YY
24/04/2026
02096316-0013